CAN ROBOTS REPLACE HUMAN JOBS

JOHN LOK

First Printing: , April 2020

Contents

Preface

Introduction

Artificial intelligence had been developed to be applied to any service , working aspects, e.g. auto-driven cars. In the future, it may replace manual transport drivers in possible, e.g. non-manual driven tram, train, bus, taxi etc. public transport service. In factory warehouse environment, it can replace some workers to deliver any goods in warehouse. In shopping center service environment, it can replace security to do patrol jobs. In restaurant, it can replace waiter to deliver food to client's table. So, AI will assist or replace any service workers to do any kinds of simple jobs in any working environment in possible future.

In my this book, I shall concenrate on discussing whether artificial intelligence can bring what advantages or disadvantages to any kinds of office environment. How can they assist office workers' tasks ? Why employers need artificial intelligence either assist or replace any office workers' tasks? Can AI only bring advantages to office working environment ? I shall indicate different kinds of industry office environment whether AI can bring only advantages or advantages and disadvantages to any kinds of office working envirnments. Readers can have more clear understanding whether AI is real suitable technologic tool to assist any office workers' tasks.

Prologue

Table of content

and agriculture and learning and medical delivery development

p.111-128

- Emergency Response to developing countries‘ earthquake natural damage suddence occurrence predicting
- Smart AI Agriculture
- Medicine Delivery to developing countries’ patients urgent need
- Assistance to reduce teaching work workload or psychological pressure to teachers in developing countries' schools
- Why does smart phone help developing countries communication ?
- The Positive Impact of Mass Media in Developing Countries
- Why do developed countries need to develop AI
- AI may bring what benefits to developed countries
- How can AI be dangerous when developed countries continue to develop AI to become weapon to replace soldiers?
- Why the recent interest in AI safety ?
- Why do developed countries people need AI ?

1

Artificial intelligence bank service working environment

Focus on outcomes not technology.Artificial Intelligence: Waiting to be unleashed? The Insider Column - When Digital Transformation misses. Are you meeting the demands of the new digital consumer? Will your legacy mindset compromise your digital competitiveness? Can artificial intelligence create online remote office new business service market in global ?

The Future of Artificial Intelligence In The Workplace:

Is AI going to displace workers or come as a benefit to them?

Is AI going to displace workers or come as a benefit to them?

Getty

Smart technologies aren't just changing our homes; they're edging their way into their numerous industries and are disrupting the workplace. Artificial Intelligence (AI) has the potential to improve productivity, efficiency and accuracy across an organization – but is this entirely beneficial?

Many fear that the rise of AI will lead to machines and robots replacing human workers and view this progression in technology as threat rather than a tool to better ourselves.

With AI continuing to be a prominent online office service business to replace human actual office working environment, businesses need to realize that self-learning and black-box capabilities are not the panacea. Many organisations are already beginning to see the incredible capabilities of AI, using these advantages to enhance human intelligence and gain real value from their data. As there is increasing evidence demonstrating the benefits of intelligent systems, more decision-makers in the boardroom are gaining a better understanding of what AI can really offer. Research conducted by EY explains "organizations enabling AI at the enterprise level are increasing operational efficiency, making faster, more informed decisions and innovating new products and services." Can articial intelligent technology create remote office working environment to replace our traditional actual office work environment ? Can we do not need to go to office to work , when any office staffs ,e.g. managers, clerk, etc. they can apply artificial intelligent technology and online technology to work at home, such as remote office working environment ?

The first companies employing AI systems across the board will gain competitive advantage, reduce cost of operations and remove head counts. Whilst this may be a positive from a business perspective, it is obvious why this a worry for those working in roles at risk of displacement. The introduction of these technologies will likely trigger an issue with unions and job security due to the substantial operational changes. Although AI will affect every sector in some way, not every job is at equal risk. PwC predicts a relatively low displacement of jobs (around 3%) in the first wave of automation, but this could dramatically increase up to 30% by the mid-2030's. Occupations within the transport industry could potentially be at much greater risk, whereas jobs requiring social, emotional and literary abilities are at the lowest risk of displacement.

A positive future with artificial intelligence to bring remote online office working environment chance:

Many businesses and individuals are optimistic that this AI-driven shift in the workplace will result in more jobs being created than lost. As we develop innovative technologies, AI will have a positive impact on our economy by creating jobs that require the skill set to implement new systems. 80% of respondents in the EY survey said it was the lack of these skills that was the biggest challenge when employing AI programs.

It is likely that artificial intelligence will soon replace jobs involving repetitive or basic problem-solving tasks, and even go beyond current human capability. AI systems will be making decisions instead of humans in industrial settings, customer service roles and within financial institutions. Automated decisioning will be responsible for tasks such as approving loans, deciding whether a customer should be onboarded or identifying corruption and financial crime.

Organisations will benefit from an increase in productivity as a result of greater automation, meaning more revenue will generated. This thus provides additional money to spend on supporting jobs in the services sector.Due to the vast array of jobs that could be impacted by AI, it is fundamental to address the potential pitfalls of these technologies. Business need to overcome the trust and bias issues surrounding AI by achieving an effective and successful implementation that makes it possible for everyone to benefit.

Governments must ensure that gains from AI are shared widely across society to prevent social inequality between those affected and unaffected by these developments. For example, this could be through increased investment into training. With the additional cost-savings from implementing AI systems, employers should also focus on upskilling their current employees.

To properly leverage the power of AI, we need to address the issue at an educational level, as well as in business. Education systems needs to focus on training students in roles directly associated to working with AI, including programmers and data analysts. This requires more emphasis to be put on STEM subjects (science, technology, engineering and mathematics). Also, subjects centered around building creative, social and emotional skills should be encouraged. Whilst artificial intelligence will be more productive than human workers for repetitive tasks, humans will always outperform machines in jobs requiring relationship-building and imagination. Hence, artificial intelligence will change our world both inside and outside the workplace. Instead of focusing on the fear surrounding automation, businesses need to embrace these new technologies to ensure they implement the most effective AI systems to enhance and compliment human intelligence.

Artificial Intelligence (AI) in Banking working environment

Artificial Intelligence (AI) is a fast-evolving technology, gaining popularity all around the world. Several industries have already adopted AI for various applications, getting better and smarter day by day. In the past few years, the banking sector has also become one of the leading adopters of Artificial Intelligence. Most banks and financial institutions are implementing AI to add more efficiency to their back-office and lessen security risks.

As per Statista, the AI market in the United States is forecasted to reach 7.35 billion U.S. dollars in 2018. Some major applications of AI include classification, image recognition, object identification, and automated geophysical feature detection. Speaking of banking and financial institutions, JPMorgan Chase, Wells Fargo, Bank of America, CitiBank, and other leading U.S. banks have already implemented AI in their systems, helping consumers manage their daily banking needs more efficiently.

AI technology can bring better Customer Support in bank service environment

Several pieces of evidence advocate that the customers willingly prefer self-service options which allow them to chat with a virtual assistant as if it were a live customer representative. Most leading banks have already added virtual assistants to their instant website chatbots, voice response systems, and mobile applications. Artificial Intelligence considers each interaction as a teachable moment, so the chatbots (virtual assistants) keeps getting better while understanding customers. With AI, virtual assistants can deliver better customer support. It also allows sentiment analysis, so the virtual assistant can determine when individuals are getting frustrated and instantly transfer them to a live agent.

Enhanced Banking Services

AI streamlines the banking process while giving customer service a new level of comfortability. It allows banks to meet customers' expectations with comprehensive digital support. With Artificial Intelligence, you can achieve greater precision and accuracy. From cash transfer to bills payment, cards management, and other support, AI can significantly enrich the satisfaction level of your customers. All of these operations can be easily managed through desktops, smartphones, and other mobile devices.

Scam Recognition

With an immense growth of banking fraud, scam recognition and reduction has become challenging for the banking sector. Several banks tried to identify the factors and powerful solutions but couldn't succeed. However, AI makes it easier to detect the factors involved in frauds and support investigators. It improves financial security with advanced fraud prevention tactics. Artificial Intelligence works as a real-time scam solution for the banking sector while handling complex situations and tactics. Based on advanced data crunching, AI can detect fraud by flagging unusual transactions. It also feeds back into the consumer's profile which subsequently builds a secure environment.

Advanced Data Analytics

One of the main advantages of AI is its ability to complete tedious tasks through intricate automation, resulting in better productivity. Based on a machine learning algorithm, AI can quickly consume and process a massive amount of data at an expedited level. The enormous speed brings efficiency to financial services, providing scope for personalized offerings to consumers. What's even more, AI makes faster decisions while carrying out actions quickly. With such advantages, it is nearly obvious that the majority of banks and financial institutions will adopt AI to stay competitive and deliver better customer support. However, several cons are also associated with a machine learning algorithm. As it continues to learn and grow, the decision-making capabilities may create problems in the near future.

Disadvanages of AI in Banking Sector

Artificial intelligence is also expected to massively disrupt banks and traditional financial services. Some of its disadvantages are listed below.

Highly Expensive

Production and maintenance of artificial intelligence demand huge costs since they are very complex machines. AI also consists of advanced software programs which require regular updates to meet the needs of the changing environment. In the case of critical failures, the procedure to reinstate the system and recover lost codes may require enormous time and cost.

Bad Calls

Though Artificial Intelligence can learn and improve, it still can't make judgment calls. Humans can take individual circumstances and judgment calls into account when making decisions, something that AI might never be able to do. Replacing adaptive human behavior with AI may cause irrational behavior within ecosystems of humans and things.

Distribution of Power

There is a constant fear of AI superseding or taking over the humans. Artificial intelligence can give a lot of power to the few individuals who are controlling it. Hence, AI carries the risk and takes control away from humans while dehumanizing actions in several ways.

Unemployment

Replacement of the workforce with machines can lead to wide-reaching unemployment. Moreover, if the use of AI becomes rampant, people will be highly dependent on the machines and lose their creative power. Unemployment is a socially undesirable issue. Individuals with nothing to do can lead to the devastating use of their minds. Be it banking or any other sector; Artificial intelligence can effectively increase the unemployment rate.

Artificial Intelligence delivered to wrong hands can turn out to be a serious threat to humankind. If individuals start thinking destructively, they can generate havoc with these advanced machines. The challenges introduced by the emergence of artificial intelligence revolve around several things. However, AI is a right balance of skill and emotions which is continually growing. Artificial intelligence provides banks, financial institutions, and tech companies with significant competitive advantages. Nevertheless, it can completely transform the financial sector and make it faster, but this will only be possible if the financial industry can manage the security risk of systems based on AI.

What does artificial intelligence mean for the bank service office workers?

With all these new artificial intelligence use cases comes the question of whether machines will force humans into obsolescence. The jury is still out: Some experts vehemently deny that artificial intelligence will automate so many jobs that millions of people find themselves unemployed, while other experts see it as a pressing problem.

"The structure of the workforce is changing, but I don't think artificial intelligence is essentially replacing jobs in bank service working environment. It allows us to really create a knowledge-based economy and leverage that to create better automation for a better form of life. It might be a little bit theoretical, but I think if you have to worry about artificial intelligence and robots replacing some bank service jobs, e.g. bank security, bank enquiry service,. But, AI can not replace bank counter service staffs to do saving or withdrawing money transfer tasks when any customers prepare to save money or withdraw money in bank counters. As this technology develops, the AI bank service will see new startups, numerous saving or withdraw transactions from consumer won't be raise more easily.

AI to Banking and Finance industry

The banking and finance industry plays a major role in our lives. I mean the world runs on money and banks are essentially the gatekeepers that regulate that flow. Did you know that the banking and finance industry heavily relies on artificial intelligence for things like customer service, fraud protection, investment, and more? A simple example is the automated emails that you receive from banks whenever you do an out of the ordinary transaction. Well, that's AI watching over your account and trying to warn you of any fraud.

AI is also being trained to look at large samples of fraud data and find a pattern so that you can be warned before it happens to you. Also, when you hitch a little snag and chat with bank's customer service, chances are that you are chatting with an AI bot. Even the big players in the finance industry use AI to analyze data to find the best avenues to invest money so they can get the most returns with the least risk. That's not all, AI is poised to play an even bigger role in the industry as major banks across the world are investing billions of dollars in the AI technology and we all will observe its effects sooner than later

2

How AI influences our daily working life in any office working environment

Can AI bring only disadvantages? If AI can bring disadvantges, what are its disadvantages to any working environment ?The entire tech world is debating the consequences of artificial intelligence and the part AI is going to play in shaping our future. While we might think that artificial intelligence is at least a few years away from causing any considerable effects on our lives, the fact remains that it is already having an enormous impact on us. Artificial intelligence is affecting our decisions and our lifestyles every day. Don't believe me? I shall indicate some product examples how AI anticipates which can influence our working culture in any office environment.

Examples of how Artificial Intelligence assistance to office working environment may include as below:

1. Smartphones

Smartphones have become the most indispensable tech product that we own today and we use it almost all the time. Well, if you are using a smartphone, you are interacting with AI whether you know it or not. From the obvious AI features such as the built-in smart assistants to not so obvious ones such as the portrait mode in the camera, AI is impacting our lives in every day office working environment.

In fact, the two examples that I provided that our working world of AI and how it is effecting our working lives. Firstly, there are the obvious AI elements which most of us have some knowledge about. For example, when you are using a smart assistant in office, whether it's Google Assistant, Alexa, Siri, or Bixby, you more or less know that these assistants are based on AI. However, when we are using a feature such as the portrait mode effect while shooting a picture, we never consider that AI might be behind that too. Have you ever thought how the Google Pixel phones or iPhones can capture such great portrait shots? The answer is artificial intelligence. So, when any office workers need to find any knowledge to solve their working problem immediately in any offices. They may apply AI smart phone tools to help them to apply online channel to search any new knowledge to attempt to solve their working problem in possible, when their computers have none any computers in offices.

Now more and more manufacturers are including AI in their smartphones with big chip manufacturers including Qualcomm and Huawei producing chips with built-in AI capabilities. The AI integration is helping in bringing features like scene detection, mixed and virtual reality elements, and more. AI is going to play an even major role in the coming years. We are already seeing the huge emphasis on AI with the latest Android and iOS updates. Features like app actions, splices, and adaptive battery in Android Pie and Siri shortcut and Siri suggestions in iOS 12 are made possible with AI. So, next time if any office workers think AI is not effecting them, take out your smartphone to replace computers to find any knowledge to help you to solve any tasks problems immediately in offices.

2. Social Media Feeds

If you are thinking that smart cars don't personally effect you as they are still not in your country or city, well, how about something which you use on a daily basis. Even if you are living under a rock, there's a high probability that you are tweeting from underneath it. If Twitter's not your choice of poison, maybe it's Facebook or Instagram, or Snapchat or any of the myriad of social media apps out there. Well, if you are using social media, most of your decisions are being impacted by artificial intelligence. So, any office workers may apply AI to help them to gather any new information to solve any difficult task problems , if their managers can not assist them to solve any sudden tasks problem, they are encountering to need to solve any working complex tasks problem internet social media in any any office working environment immediately.

From the feeds that office staffs can see in their working timeline to the notifications that you receive from these apps, everything is curated by AI. AI takes all your past behavior, web searches, interactions, and everything else that you do when you are on these websites and tailors the experience just for you. The sole purpose of AI here is to make the apps so addictive that you come back to them again and again, and I am ready to place a bet that AI is winning this war against you.

3. Online Ads Network

One of the biggest users of artificial intelligence is the online ad industry which uses AI to not only track user statistics but also serve us ads based on those statistics. Without AI, the online ad industry will just fail as it would show random ads to users with no connection to their preferences what so ever. AI has become so successful in determining our interests and serving us ads that the global digital ad industry has crossed 250 billion US dollars with the industry projected to cross the 300 billion mark in 2019. So next time when any product developers are going online and seeing ads or product recommendation, know that AI is impacting to any new products advertisement method more efficiently.

4. AI can be any office security

While we can all debate the ethics of using a broad surveillance system, there's no denying the fact that it is being used and AI is playing a big part in that. It is not possible for humans to keep monitoring multiple monitors with feeds from hundreds if not thousands of cameras at the same time, and hence, using AI makes perfect sense. With technologies like object recognition and facial recognition getting better and better every day, it won't be long when all the security camera feeds are being monitored by an AI and not a human. While there's still time before AI can be fully implemented such as security in any offices, this is going to be our future.

5. Smart Keyboard Apps

Smart Keyboard Apps. Granted, not everyone loves dealing with on-screen keyboards. However, they have become far more intuitive, allowing users to type comfortably and faster. What has probably proved to be a catalyst for them is the integration of AI. The smart keyboard apps keep a tab on the writing style of a user and predict words and emojis accordingly. Thus, typing on the touchscreen has become faster and more convenient. Not to mention, artificial intelligence also plays a vital role in pin-pointing misspellings and typos. So, any office workers can apply smart keyboard apps to help their to raise typing efficiency and reduce wrong typing word in error when they need to type any document in offices.

6. E-Commerce

` AI-driven algorithms have kind of given the much-needed impetus to e-commerce to provide a more personalized experience. According to several reports, its usage has vastly increased sales and also played a good part in building loyal relationships with customers. Thus, companies take advantage of AI to deploy chatbots to collect pivotal data and also predict purchases to create a customer-centric experience. Yet to come across this shift of strategy? Just spend some time with sites like Amazon and eBay and you will soon get to know how fast the landscape is changing around you – for the better! So, Ai can help any businesses to achieve e-commerce sale channel more easily.

7. Smart Email Apps

In any office working environment, if you still find your inbox cluttered with too many unwanted messages, chances are pretty high that you are still stuck with an old school email app. You heard it right! Modern email apps like Spark make the most of AI to get rid of spam messages and also categorize emails so that you can quickly access the important ones. What's more, they also offer smart replies based on the messages you receive to help you reply to any email quickly. The "Smart Reply" feature of Gmail is a great example of this. It uses AI to scan the text of the email and provides you with contextual answers. So, AI can help any office staffs to know who had sent any message from email and respond their email immediate , when AI can help any offices to avoid to receive any email spam rubbish email message in any time, even after working hours, it means that AI is working to help any office staffs to avoid to receive any email spam rubblish message in any time. So, when they go to office to work, even they go home after working hours. They can know whether what the important email messages are sent to their office email in boxes any time. Then, they can send email to respond their customers‘ enquires any time. So, AI can help any office workers can have chance to work at homes.

The Future of Artificial Intelligence In The Workplace Smart technologies aren’t just changing our homes; they’re edging their way into their numerous industries and are disrupting the workplace. Artificial Intelligence (AI) has the potential to improve productivity, efficiency and accuracy across an organization – but is this entirely beneficial? Many fear that the rise of AI will lead to machines and robots replacing human workers and view this progression in technology as threat rather than a tool to better ourselves.

With AI continuing to be a prominent buzzword in 2019, businesses need to realize that self-learning and black-box capabilities are not the panacea. Many organisations are already beginning to see the incredible capabilities of AI, using these advantages to enhance human intelligence and gain real value from their data. As there is increasing evidence demonstrating the benefits of intelligent systems, more decision-makers in the boardroom are gaining a better understanding of what AI can really offer. Research conducted by EY explains “organizations enabling AI at the enterprise level are increasing operational efficiency, making faster, more informed decisions and innovating new products and services."

Today In: Cybersecurity

The first companies employing AI systems across the board will gain competitive advantage, reduce cost of operations and remove head counts. Whilst this may be a positive from a business perspective, it is obvious why this a worry for those working in roles at risk of displacement. The introduction of these technologies will likely trigger an issue with unions and job security due to the substantial operational changes. Although AI will affect every sector in some way, not every job is at equal risk. PwC predicts a relatively low displacement of jobs (around 3%) in the first wave of automation, but this could dramatically increase up to 30% by the mid-2030's. Occupations within the transport industry could potentially be at much greater risk, whereas jobs requiring social, emotional and literary abilities are at the lowest risk of displacement.

A positive future with artificial intelligence

Many businesses and individuals are optimistic that this AI-driven shift in the workplace will result in more jobs being created than lost. As we develop innovative technologies, AI will have a positive impact on our economy by creating jobs that require the skill set to implement new systems. 80% of respondents in the EY survey said it was the lack of these skills that was the biggest challenge when employing AI programs. It is likely that artificial intelligence will soon replace jobs involving repetitive or basic problem-solving tasks, and even go beyond current human capability. AI systems will be making decisions instead of humans in industrial settings, customer service roles and within financial institutions. Automated decisioning will be responsible for tasks such as approving loans, deciding whether a customer should be onboarded or identifying corruption and financial crime. Organisations will benefit from an increase in productivity as a result of greater automation, meaning more revenue will generated. This thus provides additional money to spend on supporting jobs in the services sector.

How to take advantage of AI to any offices

Due to the vast array of jobs that could be impacted by AI, it is fundamental to address the potential pitfalls of these technologies. Business need to overcome the trust and bias issues surrounding AI by achieving an effective and successful implementation that makes it possible for everyone to benefit. Governments must ensure that gains from AI are shared widely across society to prevent social inequality between those affected and unaffected by these developments. For example, this could be through increased investment into training.With the additional cost-savings from implementing AI systems, employers should also focus on upskilling their current employees.

To properly leverage the power of AI, we need to address the issue at an educational level, as well as in business. Education systems needs to focus on training students in roles directly associated to working with AI, including programmers and data analysts. This requires more emphasis to be put on STEM subjects (science, technology, engineering and mathematics). Also, subjects centered around building creative, social and emotional skills should be encouraged. Whilst artificial intelligence will be more productive than human workers for repetitive tasks, humans will always outperform machines in jobs requiring relationship-building and imagination. Artificial intelligence will change our world both inside and outside the workplace. Instead of focusing on the fear surrounding automation, businesses need to embrace these new technologies to ensure they implement the most effective AI systems to enhance and compliment human intelligence

How AI can help office workers to do tasks more easily

Companies are currently spending big on artificial intelligence and machine learning initiatives to the tune of $12 billion, but estimates put that figure as high as $57.6 billion by 2021, according to the International Data Corporation (IDC). With such massive shifts, the focus is usually on what we might lose, but it shouldn't be. A recent report on the future of work from the McKinsey Global Institute suggests that while only about 5% of jobs can be completely eliminated by automation, the rise of AI requires workers to beef up both technical and soft skills in order to stay competitive.

What's seldom discussed is how AI can revolutionize our jobs. It's now possible to pinpoint peak productivity for a single day, improve communication in meetings (even before people ever work together face to face), or even teach you to be a better leader, all thanks to AI platforms. I shall indicate these advantages to bring any office benefits from AI assistance as below:

1. AI can help any companies to get better to hire the best applicants

AI has the greatest potential to change the way companies find candidates, according to Alexander Rinke, cofounder and CEO of Celonis. The company's process-mining technology helps businesses to understand the areas where automation can help humans, he says. In HR departments, Celonis can help identify how fast workers come and go, the cost per hire, and which positions take the longest to fill. AI helped enable one customer's ability to identify bottlenecks

in recruitment and reduced process costs internally by 30% as well as get them hired more quickly, he says.
Crafting a resume has never been easier, nor has landing an interview. Another example is how recruitment software provider iCIMS, in partnership with Google, is helping job seekers find jobs directly through the search engine, thanks to Google's AI and machine learning capabilities. Susan Vitale, iCIMS's chief marketing officer says that in addition to reducing the number of expired job postings, machine learning is underlying a private beta program of Google's Cloud Jobs Discovery model. "For a candidate searching for, say, a CTO role, Cloud Job Discovery will serve up CTO positions as well as jobs with titles that are similar, but not verbatim, such as chief technology officer or chief technical officer," says Vitale. This model also allows for conceptual search results, such as serving up job listings for cashiers, sales associates, and store associates when someone searches for one versus just only showing jobs that exactly match the keyword search criteria, she adds.

2. AI can help any office workers to raise much more productive efficiencies

John Furneaux, CEO and cofounder of Hive, says predictive analytics will help us better understand how we work. "It can tell us just about everything we want to know about teams and collaboration, for example, if men or women get more done in the afternoon, and if summer Fridays are a myth," he says. (Everyone thinks summer Fridays aren't productive, but in reality there's no difference between those and other Fridays during the year–productivity is equally low.)
Using a data set of over 30,000 completed actions across Hive workspaces, Furneaux says they were able to identify some notable trends in productivity. For example, men were

far more productive early in the day, with a sharp decline in the afternoon, while women had a slower start to the day but were far more productive in later hours than their male counterparts. And analyzing chat messages revealed that women appear to complete more tasks when chatting, suggesting they use communication as a key tool to completing work. Similarly, Nintex Hawkeye analyzes data on business processes by types, users, roles, and departments to see who's doing the work and how long it takes them to do it. Management can monitor and analyze those metrics in real time.

3. AI can help any managers to make the most fair compensation and eliminate wage gaps to every staffs

Tanya Jansen, cofounder of the compensation management platform beqom, says that AI and predictive analytics can eliminate unconscious bias from compensation. Jansen says that AI based on a variety of rules including education, experience, certifications, and more can make compensation more fair and help businesses move closer to closing pay gaps. "Specifically, AI can help solve gender pay gaps and the CEO-to-worker pay gap, in which pay ratios of Fortune 500 companies range from 2:1 at the low end to nearly 5000:1 at the high end," she says. Additionally, the use of AI-driven compensation technology to make pay more fair can mitigate the risk of employee turnover, which costs businesses as much as 33% of a worker's annual salary to replace them.

4. AI can help any office staffs to arrange better meetings

Augmented Reality (AR) is still in its infancy, but AI and machine learning are the core components that make it work. As such, Christa Manning, the vice president and solution provider research leader at Bersin, Deloitte Consulting LLP, says that AR can help workers find the right information, in the right place, at the right time to make the best decisions wherever they may be working. For example, as more companies adopt video meetings and collaborative workspaces, it's likely we'll begin to see HR-curated information like talent profiles and work styles layered over interactions through AR."Imagine being in a video conference with a colleague and having direct insight into their communication style, seeing tips on how to best interact with them or reminders of what needs to be discussed. SO, AI can help any organizations to conclude or find the best methods to solve any problems after their every discussion in any meetings.

How AI is improving onboarding and training. AI coaching tools first learn by observing how different employees conduct specific tasks. Then these tools can walk new employees through how to complete those tasks—or even coach existing employees on how to do things more effectively or efficiently. Chorus is a great example of this technology. It analyzes sales calls while they happen, offering tips to help sales reps manage the cadence of meetings and use the most effective messaging. It also records all sales calls and compiles statistics for each sales rep, providing everyone with the tools they need to help them close more deals and conduct more effective calls. Another example is Cogito, a tool that combines AI with behavioral science to help customer service employees provide better phone support. It monitors calls for voice signals, providing real-time suggestions to representatives on how to improve the conversation.

5. AI can help any managers to be better leaders

Indiggo, a platform powered by a proprietary AI tool called "indi," functions as a brain that has consumed all the knowledge the company has gathered in its 15 years of operation. It also uses an algorithm to provide an estimate of how much time is wasted by a company by analyzing the size of its management team. Then it taps their calendars to see how they spend their time, and walks individual managers through a type of Q&A to make sure they are clear on what their top three priorities are, and how that relates to the organization's priorities, which will indicate if that strategy is moving forward or not. "The counterintuitive impact of these advances is that they actually make human work truly irreplaceable," Alexander Rinke, the cofounder and CEO of Celonis says. As such, he reminds us, "Humans are much better at processes that involve reasoning, judgment, and interaction with people." So, AI can recommend more accurate and useful opinions to help any managers to solve their managing challenges in office any time.

How AI is eliminating repetitive administrative tasks

There are a lot of tasks that knowledge workers spend time on that provide little—if any—value.For example, say you need to schedule a meeting to get consensus on a decision before moving forward, but you need five people to join the meeting. It's easy to spend a ton of time sending email back-and-forth or finding an open slot on everyone's calendar.That's not the most rewarding use of your time for you or your company.Tools like X.ai give employees AI-powered personal assistants that perform administrative tasks like scheduling, rescheduling, and cancelling meetings.

How AI is transforming internal communications and support

Personnel on the teams that provide employee support have their hands full with other responsibilities, too. HR teams work on building the kind of company people love working for. IT maintains the company's network and keeps data secure. Office managers frequently run big events like holiday parties.These tasks are crucial, but they're often hard for teams to focus on because they're busy answering routine questions. AI service desks like askSpoke allow employee support teams to balance their service commitments with other important responsibilities by reducing interruptions from rote, repetitive requests.Employees can askSpoke for whatever they need over Slack, email, SMS, and the web. askSpoke's friendly AI will automatically provide a prompt response.

How AI is transforming marketing, sales, and customer service

AI-powered chatbots help with external support as well. Just like with internal support tools like askSpoke, these chatbots learn from real marketers, salespeople, and customer service reps and are eventually able to answer questions as accurately as a knowledgeable person.For example, chatbot for Messenger helps customers plan their vacations. It books flights, hotels, and cars, highlights destination attractions, and even provides answers to questions like "Where can I go for $100 expense budget only?"

How AI is transforming business data and analytics

It's hard to run a competitive business today without data. But even massive amounts of data are useless without a way to transform that data into valuable insights. That's typically why you'd want to hire a data scientist—which just happens to be one of the most difficult roles to fill. How AI is fighting fraud and transforming security. Have you ever taken a call from your bank to find that someone used your debit card fraudulently? Most likely, your bank used some form of AI to detect the fraudulent transaction and decline it. Applying the same basic technology to the workplace helps identify security risks and keeps customer, employee, and company data safe. AI-powered software can automatically detect and address threats among thousands or millions of signals that humans would never be able to parse (especially not in real-time).

How AI is transforming productivity

While AI is transforming the workplace in many different ways across every industry, it's impacting productivity most of all. When your office staffs don't have to scroll through calendars to look for open meeting times, build reports in spreadsheets to look for insights, or spend your day answering the same questions over and over again, you're more productive. Workers are freed from redundant and mindless tasks, giving them more time to do work that matters, solve problems, and exercise their creativity. Some tools use AI to specifically monitor and boost productivity. For example, Deloitte's LaborWise provides company leaders and managers with productivity analytics that help them identify areas where labor costs are too high, impediments that slow people down, and departments that need additional staff.

In conclusion, what AI means for the workplace of the future. While some will dramatize the negative impacts of AI, cognitive computing, and robotics, these powerful tools will also help create new jobs, boost productivity, and allow workers to focus on the human aspects of work. Essentially, automation frees companies and their employees up to be more empathetic, to focus on things like the customer experience, employee engagement, and workplace culture.

What are traditional office tools to be replaced by AI ?

Artificial intelligence (AI) is predicted to eliminate over a million jobs in the next few years, potentially replacing lower level positions like administrative assistants with humanoid robots or voice assistants. But in the nearer future, fresh AI-driven software and products are also moving to eliminate non-human elements of the workplace by replacing traditional office tools, including both physical products and everyday electronic processes. Why should businesses switch from the tried-and-true to emerging technology? Many of the experts TechRepublic talked to said the AI options streamline business practices, making their adopters work smarter instead of harder. I shall indicate these office tools ,they can be applied to help any office staffs to finish their these tasks in office, they may include as below:

1. Scheduling

Workloud's end-to-end, cloud-based workforce management software takes scheduling from paper or Excel and moves it to the cloud. Everything from clocking in and out to monitoring employee absences is fully digitalized.Schedules and timesheets are accurate, created easily, and accessible through the service's web, tablet, and mobile apps. The software can also be used for absence management.

2. Employee talent selection

Using AI and organizational behavior science, can be used to replace internal spreadsheets and databases designed to monitor human capital. By mining employee attributes and experiences, the software can recommend who would be best for a project. The software also collects reviews after projects to better predict successful employee-project matches.The traditional hiring process is slow, biased and inaccurate, By removing humans from the beginning stages of the process, it can become faster and more fair, and result in better hires.

AI software automates the hiring process, using online simulations instead of manual screenings and interviews. Using the software, employers can include tasks in a job application, allowing job candidates to show technical skills that may be necessary for a job. Employers can't rule out candidates until they see how the candidate performs, eliminating bias that occurs in the resume reading stage. Both sides also automatically receive updates about each other's steps, reducing the amount of time it takes to .

3.Timesheets: Allocate

Using AI and machine learning, the software registers an employee's computer activity throughout the day. The data, which can also pull information from email and calendars, is used to suggest timesheet entries to reflect a more accurate amount of time an employee spent working. The employee can review and revise as necessary. However, the software doesn't spy on or monitor employees. The data is only available to each employee, while others in the company can only see the timesheet's output, which Allocate said would be the same information available if a manual sheet was used. So,replacing manual timesheets with Allocate has three advantages: More accurate time entry, project analytics, and "'unsucking' the work experience."

4. Document storage

By using AI to read and analyze business and legal documents, AI can store all of the important document-based information in the cloud. The severe reduction in print-outs means less paper and ink, fewer products like binder clips and boxes to store and organize all of the paper, and more employee time freed up from not needing to manually sort through every document.

For example, in any lawyer offices, legal professionals'

morale in the industry can suffer when they are pushed into performing such dull, repetitive tasks like sorting through and coding documents by hand, With AI tools to automate those duties, lawyers can focus on more meaningful projects and boost the business's and clients' success as a result. While focused on law firms, businesses that have a lot of unstructured data in documents may also be able to use the service to free up employee time and save on printing costs.

5. Scanners: Adobe Scan

While documents are moving to the cloud more and more, sometimes a physical copy of a document still needs to be scanned using a bulky office scanner. Adobe Scan, an app that condenses a scanner to the size of a smartphone, can rid offices of the need for an in-house scanner. Users can download and open the app, then hold their device over whatever they need to scan. Adobe Sensei then turns the scan into a PDF, and sends it to the Adobe Document Cloud. The app can transform any image into digital text that can then be searched and used electronically. The app streamlines the scanning process, making scans cleaner and more immediate. For businesses already using Adobe services, the app makes documents easily accessible.

6. Landline phones

While landlines in homes are increasingly less common, the same cannot be said for offices. But using chatbots and AI integrations, RingCentral is trying to replace traditional office landline phone systems. The platform offers over 100 integrations, including that AI landline phones can let employees check their voicemail, and a Gong.io option that listens to call recordings to find traits of successful employees than can be used in training. An add-on for Gmail lets users switch from emailing back and forth to a voice session without needing to look up contact information. AI landline phone is easy to adopt and use in the workplace, and is more customizable than standard phone systems, said David Lee, vice president of platform products. Compared to the traditional option, the cloud-based option is "future-proof.

3

How artificial intelligence can raise office efficiency

Artificial Intelligence is already impacting every industry through automation and machine learning, bringing concerns that AI is on the fast track to replacing many jobs. But these fears aren't new, says Dan Jackson, director of Enterprise Technology at Crestron, a company that designs workplace technology. "I'd argue this is no different than when we moved from an agricultural to an industrial economy at the turn of the last century. The percentage of people working in agriculture significantly decreased, and it was a big shift, but we still have plenty of jobs 100 years later," he says. Anytime society experiences a major technological advancement, we need to be prepared for it to change the way we live and work. It's hard to imagine what the future of jobs will look like with AI, but that future exists. And optimists suggest that, like the sewing machine to the textile industry, AI will make us better, more efficient and faster workers.

In fact, many experts agree that AI has the potential to

eliminate mundane, administrative work, while we will always rely on human workers to be empathetic, collaborative, creative and strategic. But it's impact on any industry lies in the hands of the business leaders who are responsible for adopting AI strategies.

- Training presents challenges

A recent study of 1,000 global companies by Accenture found that AI is already creating three new categories of jobs: trainers, explainers and sustainers. Trainers are the people who teach AI systems how to act -- whether it's language, human behavior or the intricacies of human interaction. Explainers are the liaison between technology and business leaders, providing more insight and clarity into machine learning for the non-tech workers. Sustainers are the workers required to maintain AI systems and troubleshoot any potential issues. Some jobs were highly technical and required advanced degrees, but other roles demanded innately human things such as empathy and interaction. Downstream jobs, such as those in sales, marketing, or service will change to take advantage of the insights from AI, but many of the core skills will remain. However, it might sound like any job related to AI will require years of technical knowledge, but that isn't the case. We've already seen a shift in tech hiring -- companies often need highly specific skill sets that are hard to find in potential candidates. As a result, more businesses are hiring employees with the right soft skills, and then training them in technical skills.

An office effort measured approach to AI

The real takeaway is that any approach to AI will need to consider the human aspect of every business. AI has great potential to increase efficiency and accuracy and it's already been proven in certain industries. For example, the use of AI In banking to identify and money laundering schemes. It's also improved healthcare by "increasing the speed and accuracy" of cancer diagnosistics. AI can also help reduce the cost and length of human trafficking investigations, a situation where time is precious. In these examples, AI hasn't replaced jobs, but has positively impacted efficiency.

Thus, we need to ensure our education system responds to equip young people with the appropriate skills and adaptability, while businesses and public organizations must invest in training. Perhaps most of all, we need to encourage imagination and willingness to experiment. The organizations that can innovate with AI will reap the benefits. Their growth will make them the primary source of future jobs. Companies have a choice when implementing AI. They can choose to effectively implement systems that make employee's lives easier and find creative ways to leverage the technology. It's up to employers to ease fears for workers around AI and build strategies that benefit everyone. Hence, some AI experts believe AI can only raise efficiency to some office tasks, however, AI can not still raise efficiency to all office tasks for any office deparments. The reasons are because some office tasks which can only dominate to finish by human office workers. These office tasks are as below:

How can leaders and managers improve employee productivity while still saving time? These below tasks, AI experts ensure that AI can not help any office workers to raise their efficiencies as below:

1. Office managers can not delegate to AI to help them to do. While this tip might seem the most obvious, it is often the most difficult to put into practice. We get it–your company is your baby, so you want to have a direct hand

in everything that goes on with it. While there is nothing wrong with prioritizing quality (it is what makes a business successful, after all), checking over every small detail yourself rather than delegating can waste everyone's valuable time. Instead, give responsibilities to qualified employees, and trust that they will perform the tasks well. This gives your employees the opportunity to gain skills and leadership experience that will ultimately benefit your company. You hired them for a reason, now give them a chance to prove you right.

2. Office managers can not match Tasks to Skills to AI. Knowing your employees' skills and behavioral styles is essential for maximizing efficiency. For example, an extroverted, creative, out-of-the-box thinker is probably a great person to pitch ideas to clients. However, they might struggle if they are given a more rule-intensive, detail-oriented task. Asking your employees to be great at everything just isn't efficient–instead, before giving an employee an assignment, ask yourself: is this the person best suited to perform this task? If not, find someone else whose skills and styles match your needs.

3. Office managers can not teach AI to replace them how to communicate and teach their low level staffs how to work effectively. Every manager knows that communication is the key to a productive workforce. Technology has allowed us to contact each other with the mere click of a button (or should we say, tap of a touch screen)–this naturally means that current communication methods are as efficient as possible, right? Not necessarily. A McKinsey study found that emails can take up nearly 28% of an employee's time. In fact, email was revealed to be the second most time-consuming activity for workers (after their job-specific tasks). Instead of relying solely on email,

try social networking tools (such as Slack) designed for even quicker team communication. You can also encourage your employees to occasionally adopt a more antiquated form of contact...voice-to-voice communication. Having a quick meeting or phone call can settle a matter that might have taken hours of back-and-forth emails. All of above communication tasks, I believe that AI can not do better than managers in offices.

4. AI can not keep Goals Clear and focused to be better than managers. You can't expect employees to be efficient if they don't have a focused goal to aim for. If a goal is not clearly defined and actually achievable, employees will be less productive. So, try to make sure employees' assignments are as clear and narrow as possible. Let them know exactly what you expect of them, and tell them specifically what impact this assignment will have. One way to do this is to make sure your goals are "SMART" – specific, measurable, attainable, realistic, and timely. Before assigning an employee a task, ask yourself if it fits each of these requirements. If not, ask yourself how the task can be tweaked to help your workers stay focused and efficient.

5. AI can not know how to incentivize Employees to work more efficiently. One of the best ways to encourage employees to be more efficient is to actually give them a reason to do so. Recognizing your workers for a job well done will make them feel appreciated and encourage them to continue increasing their productivity. When deciding how to reward efficient employees, make sure you take into account their individual needs or preferences. For example, one employee might appreciate public recognition, while another would prefer a private "thank you." In addition

to simple words of gratitude, here are a few incentives managers can know how to incentivize their staffs to work efficiently, but AI is only one machine, it can not perform very good.

6. AI does not know how to assist managers to train and Develop employees. Reducing training, or cutting it all together, might seem like a good way to save company time and money (learning on the job is said to be an effective way to train, after all). However, this could ultimately backfire. Forcing employees to learn their jobs on the fly can be extremely inefficient.

So, instead of having workers haphazardly trying to accomplish a task with zero guidance, take the extra day to teach them the necessary skills to do their job. This way, they can set about accomplishing their tasks on their own, and your time won't be wasted down the road answering simple questions or correcting errors. Past their original training, encourage continued employee development. Helping them expand their skillsets will build a much more advanced workforce, which will benefit your company in the long run. There are a number of ways you can support employee development: individual coaching, workshops, courses, seminars, shadowing or mentoring, or even just increasing their responsibilities. Offering these opportunities will give employees additional skills that allow them to improve their efficiency and productivity. But, AI do not know how to improve any office workers' performance more easily than managers.

- How can AI be dangerous to office working environment?

Most researchers agree that a superintelligent AI is unlikely to exhibit human emotions like love or hate, and that there is no reason to expect AI to become intentionally benevolent or malevolent. Instead, when considering how AI might become a risk to any office working environments, experts think two scenarios most likely:

The AI is programmed to do something devastating: Autonomous weapons are artificial intelligence systems that are programmed to kill. In the hands of the wrong person, these weapons could easily cause mass casualties. Moreover, an AI arms race could inadvertently lead to an AI war that also results in mass casualties. To avoid being thwarted by the enemy, these weapons would be designed to be extremely difficult to simply "turn off," so humans could plausibly lose control of such a situation. This risk is one that's present even with narrow AI, but grows as levels of AI intelligence and autonomy increase. So, if some businessmen apply AI to be business weapon to attack or steal their business competitors' business secret, e.g. contract document, employee performance report, profit report, even business secret document. Then, AI will be one business competitor weapon more than business assistant role in any business market. So, whether AI is office assistant or business competitor weapon, it depends on how the businessmen apply them to assist their business development.

The AI is programmed to do something beneficial, but it develops a destructive method for achieving its goal: This can happen whenever we fail to fully align the AI's goals with ours, which is strikingly difficult. If you ask an obedient intelligent car to take you to the airport as fast as possible, it might get you there chased by helicopters and covered in vomit, doing not what you wanted but literally what you asked for. If a superintelligent system is tasked with a ambitious geoengineering project, it might wreak havoc with our ecosystem as a side effect, and view human attempts to stop it as a threat to be met.

As these examples illustrate, the concern about advanced AI isn't malevolence but competence. A super-intelligent AI will be extremely good at accomplishing its goals, and if those goals aren't aligned with ours, we have a problem. You're probably not an evil ant-hater who steps on ants out of malice, but if you're in charge of a hydroelectric green energy project and there's an anthill in the region to be flooded, too bad for the ants. A key goal of AI safety research is to never place humanity in the position of those ants. Because AI has the potential to become more intelligent than any human, we have no surefire way of predicting how it will behave. We can't use past technological developments as much of a basis because we've never created anything that has the ability to, wittingly or unwittingly, outsmart us. The best example of what we could face may be our own evolution. People now control the planet, not because we're the strongest, fastest or biggest, but because we're the smartest. If we're no longer the smartest, are we assured to remain in control?

A captivating conversation is taking place about the future of artificial intelligence and what it will/should mean for humanity. There are fascinating controversies where the world's leading experts disagree, such as: AI's future impact on the job market; if/when human-level AI will be developed; whether this will lead to an intelligence explosion; and whether this is something we should welcome or fear. But there are also many examples of of boring pseudo-controversies caused by people misunderstanding and talking past each other. To help ourselves focus on the interesting controversies and open questions — and not on the misunderstandings — let's clear up some of the most common myths.

There have been a number of surveys asking AI researchers how many years from now they think we'll have human-level AI with at least 50% probability. All these surveys have the same conclusion: the world's leading experts disagree, so we simply don't know. For example, in such a poll of the AI researchers at the 2015 Puerto Rico AI conference, the average (median) answer was by year 2045, but some researchers guessed hundreds of years or more. There's also a related myth that people who worry about AI think it's only a few years away. In fact, most people on record worrying about superhuman AI guess it's still at least decades away. But they argue that as long as we're not 100% sure that it won't happen this century, it's smart to start safety research now to prepare for the eventuality. Many of the safety problems associated with human-level AI are so hard that they may take decades to solve. So, any businessmen ought have business moralty to know whether they ought how to apply their AI to assist their business development in our future office environment to be more moral.

- Five ways to use AI to improve business efficiency to these office tasks

Regardless of a company's size or type, its executives typically look for ways to help it operate as efficiently as possible. They understand the link between efficiency and profitability. If employees waste too much time with drawn-out processes or complicated tasks, it'll be hard for the enterprise to remain profitable and adapt to challenges. Fortunately, artificial intelligence (AI) supports the need for effective business operations. Here are five ways enterprises can use AI for help: 5 ways to use AI to improve business efficiency image.Getting the best results from AI means looking at where bottlenecks exist, then figuring out if and how it might remove or minimise them. AI can help any offices to improve or raise efficiency to these tasks aspects as below:

1. Use AI to answer queries and support customer engagement

Chatbots are an increasingly popular option for businesses to try, and they use AI to work. Companies often build chatbots that can answer any questions from customers that come through outside of business hours. Some identify the nature of a person's problem, then either attempt to tackle it with preprogrammed answers or pass the communications to a human support worker. The retail industry, in particular, saw success by deploying chatbots. Global data collected by Juniper Research shows an estimated 2.6 billion retail-based chatbot interactions in 2019, and the company forecasts the number to rise to 22 billion in 2023.

Chatbots are excellent for answering simple questions like "How late are you open today?" or "Do you have gluten-free menu options?" Getting quick answers to queries like those increases the chances customers will choose to do business with one company over another. Equally importantly, when chatbots can give responses in a matter of seconds, there's no need for humans to stop what they're doing and address the questions.

2. To enhance reporting speed and accuracy

Company reports reveal things such as which products are selling the fastest and where they're most popular. They can also confirm the impacts of marketing campaigns on product sales, break down the costs of a new packaging choice or shipping method, and much more. However, as anyone that files reports knows, creating them is a painstaking task, and trying to rush through the process could cause mistakes. Some forward-thinking companies are combining AI with big data analytics. Doing this brings better forecasts and takes some of the burdens off the people who prepare the reports. AI also helps conquer the inevitability of mistakes. Even the most careful people make blunders, often because of mental fatigue.

AI learns to spot patterns in data and gets smarter with time. This means reports get finished faster and contain more-reliable information. The reliability aspect is crucial, especially since recently published research indicated two-thirds of the senior executives polled had no confidence or trust in big data. Using AI does not mean companies can do without data scientists. However, depending on the technology allows them to reduce the uncertainty that may otherwise exist. It also prevents employees who work with a company's data from being asked to recheck the findings, even if they initially took appropriate precautions to ensure accuracy.

3. To improve data transfer speeds

Fast data transfers help AI technology work. Concerning some information-intensive applications like virtual reality (VR), any slow transmissions greatly interfere with the realism, and content immersion people should enjoy after strapping on a VR headset. As it turns out, AI can improve data transfer speeds, too. For example, services exist that boost speeds across any wide-area network (WAN). Users enjoy consistently accelerated rates regardless of the kind of information transferred. Some companies have solutions that can reduce WAN job times by up to 98%. These AI-driven options work particularly well when companies need to move information between data centres or cloud environments.

4. To assist the IT team with identifying genuine cyberthreats and anomalies

One of the ongoing challenges faced by IT teams of all sizes is to separate the true cyber threats from false alarms. The difficulties associated with categorising the two types may mean cybersecurity professionals waste time getting to the bottom of things that are ultimately nonissues. They might miss the actual threats that could derail a company's operations. Besides detecting possible intrusions associated with a network, AI can screen for software abnormalities that may make it easier for cybercriminals to orchestrate their attacks successfully. It can also find malicious software hackers installed. Due to this kind of information and the advantages of receiving it through real-time updates, IT security teams can work more productively. They can use the majority of their resources on the threats that matter most to the company's stability.

Some organisations have even used AI to help them conquer the substantial skills shortage in the cybersecurity industry. At Texas A&M University, the Security Operations Center deals with about a million attempted hacks each month. The facility has some full-time workers, but students comprise most of the staff. They work alongside AI that aids in threat monitoring, detection and remediation. Before students see possible threats, the smart technology finds and groups them. This approach saves time and lets the team get to work investigating the problems and deciding how to handle them.

5. To streamline the time-to-hire metric when filling new positions

Statistics show the average time required to hire a person for an open position ranges from 12.7 to 49 days, depending on the industry. The timing also varies based on the type of work a job requires. For example, it takes a shorter amount of time overall to find someone for an administrative or human resources position than one associated with a creative or advertising role. Then, of course, interviews are more extensive for high-profile work.

Human resources professionals increasingly use AI to cut down on the time between first posting a job and finding the ideal individual to hire. For example, an AI platform could look for particular desired keywords in submitted resumes, saving hiring managers from poring over the documents themselves. AI can also pitch in during interviews. A company called VCV recently raised $1.7m to further develop its AI tool that has voice and facial recognition components. Candidates are asked to record videos of them answering interview questions, but they can't prepare for the specific content in advance.

In conclusion, AI Can Boost Efficiency at All Types of Companies. The examples here highlight why so many company leaders conclude that if they use AI, they could cut down on inefficiencies. Getting the best results from AI means looking at where bottlenecks exist, then figuring out if and how it might remove or minimise them. But, AI still lack enough effort to help all staffs to raise efficiency to all department tasks in any office environments.

- How the office energy Department is using AI to solve some of their office staffs electricity toughest challenges in their office working environment.

Insights from artifical intelligence has the potential to transform nearly every aspect of the world as we know it.

Today, it is being applied to accelerate the pace of discovery in a wide variety of areas including energy, materials science, health care, national security, emergency response, transportation, and more. AI can be trained to help any energy department to gather data to avoid energy waste to be used to any organizations. So, AI is such as one super machine to do more accurate judgement to help any energy scientists to find the best methods to help any organizations to avoid to waste to use any energy daily. Then, organizations can avoid to spend too much energy to use in offices and they can save more money and avoid energy shortage challenge causes more easily. When the office managers can apply AI ability to reason and put it into a more automated format in a computer system to their every staffs' computer and record their computer electricity use record in their offices every day.

How can AI help offices to save energy or avoid to waste energy ?

The next industrial revolution is already happening. Artificial intelligence (AI) is ushering in an era of technologies that are faster, more adaptable, more efficient, and making the world more digitally connected. AI is best described as complementary to human intelligence, delivering the computing power to crunch numbers too big for people and recognize patterns too tedious for the human eye. In a Harvard Business Review study of 1,500 companies, it was found that the most significant performance improvements were made when humans and machines worked together. As AI becomes one of society's greatest assets, it's especially helpful for solving problems that seem larger than life — like protecting our natural environment.

Through machine learning, robotics, drones, and the

internet of things (IoT), society can achieve better monitoring, understanding, and prevention of damage and stressors on Earth's land, air, and water. Even technology already available today could reduce energy usage in the U.S. by 12 to 22 percent, according to The Information Technology Industry Council (ITI). In the face of this dire reality, the potential of technology to help meet this challenge is a rare source of optimism. According to a recent survey by Intel and the research firm Concentrix, 74 percent of business-decision makers working in environmental sustainability agree artificial intelligence (AI) will help solve long-standing environmental challenges; 64 percent agree the Internet of Things (IoT) will help solve these challenges. As the field of AI develops, so will the potential to protect the environment. From the land and air to both drinking and ocean water, AI is shaping up to be the key that governments, organizations, and individuals can tap to work toward a cleaner planet, even AI can help offices to avoid to waste energy when staffs are working in offices every day.

Many AI scientists indicate that AI will also make renewable energy technology like solar panels and wind turbines more efficient and cost effective, helping them to become ubiquitous and lower society's dependence on fossil fuels. AI will also make renewable energy technology like solar panels and wind turbines more efficient and cost effective, helping them to become ubiquitous and lower society's dependence on the fossil fuels polluting the air — then hopefully eliminate them all together. Combined with the smart grid, another technology that will be enabled by AI, this will truly progress the way people receive and use electricity in their homes, offices, and everywhere else. Smart meters save energy by allowing for two-way communication between the grid and anything that uses electricity, giving energy providers a better understanding of usage and the ability to make real-time adjustments for efficiency. Customers will benefit from the real-time data too; seeing the increased costs at peak times will encourage them to voluntarily adjust their usage to save money. This will, in turn, save even more energy: a win-win. Plus, the process of delivering the energy itself will also be improved by the smart grid, thanks to Volt/VAR control systems that can reduce the amount of energy wasted when it's in electricity transmission lines.

4

Can AI replace office workers

Can AI replace all office workers to do their different tasks in office different department ? If AI can only replace some department office workers to do their simple tasks, how it can raise more efficiency to compare them in some business office environments. I shall indicate some office tasks to explain how AI can help these businesses to raise their efficiency in officesas below:

- AI insurance workers

Nowadays, some country offices begin apply robotics to replace human office workers in their companies. For example, Japanese company replaces office workers with artificial intelligence in insurance industry. A future in which human workers are replaced by machines is about to become a reality at an insurance firm in Japan, where more than 30 employees are being laid off and replaced with an artificial intelligence system that can calculate payouts to policyholders.

Fukoku Mutual Life Insurance believes it will increase productivity by 30% and see a return on its investment in less than two years. The firm said it would save about 140m yen (£1m) a year after the 200m yen (£1.4m) AI system is installed this month. Maintaining it will cost about 15m yen (£100k) a year. The move is unlikely to be welcomed, however, by 34 employees who will be made redundant by the end of March.

The system is based on IBM's Watson Explorer, which, according to the tech firm, possesses "cognitive technology that can think like a human", enabling it to "analyse and interpret all of your data, including unstructured text, images, audio and video".The technology will be able to read tens of thousands of medical certificates and factor in the length of hospital stays, medical histories and any surgical procedures before calculating payouts, according to the Mainichi Shimbun.

While the use of AI will drastically reduce the time needed to calculate Fukoku Mutual's payouts – which reportedly totalled 132,000 during the current financial year – the sums will not be paid until they have been approved by a member of staff, the newspaper said.

Japan's shrinking, ageing population, coupled with its prowess in robot technology, makes it a prime testing ground for AI. According to a 2015 report by the Nomura Research Institute, nearly half of all jobs in Japan could be performed by robots by 2035. For example, one Japan insurance company, Dai-Ichi Life Insurance has already introduced a Watson-based system to assess payments - although it has not cut staff numbers - and Japan Post Insurance is interested in introducing a similar setup, the Mainichi said. AI could soon be playing a role in the country's politics. Next month, the economy, trade and industry ministry will introduce AI on a trial basis to help civil servants draft answers for ministers during cabinet meetings and parliamentary sessions. The ministry hopes AI will help reduce the punishingly long hours bureaucrats spend preparing written answers for ministers.

- AI public service workers

The automated city: do we still need humans to run public services? If the experiment is a success, it could be adopted by other government agencies, according the Jiji news agency. If, for example a question is asked about energy-saving policies, the AI system will provide civil servants with the relevant data and a list of pertinent debating points based on past answers to similar questions.

The march of Japan's AI robots hasn't been entirely glitch-free, however. At the end of last year a team of researchers abandoned an attempt to develop a robot intelligent enough to pass the entrance exam for the prestigious Tokyo University. "AI is not good at answering the type of questions that require an ability to grasp meanings across a broad spectrum," Noriko Arai, a professor at the National Institute of Informatics, told Kyodo news agency. Hence, AI will have possible to replace some public service workers' tasks.

- AI replace warehouse workers

Denso's use of Drishti shows how some jobs will be transformed by artificial intelligence even when they're unlikely to be eliminated by AI anytime soon. Many jobs in manufacturing require dexterity and resourcefulness, for example, in ways that robots and software still can't match. But advances in AI and sensors are providing new ways to digitize manual labor. That gives managers new insights—and potentially leverage—on workers. For example,some workers say the results are unpleasant. Last year, Amazon warehouse employees in Minnesota staged a walkout to protest how the company uses inventory and worker-tracking technology. They allege that Amazon uses it to enforce a punishing working pace that causes injuries. The company has disputed those claims, saying it coaches employees on how to safely meet quotas.

Workers at Denso were initially wary of the prospect of being video-recorded all day to feed machine-learning algorithms, but Huffman says they have since come to appreciate Drishti's technology. After something goes wrong, workers can now look at the data and video with their managers, instead of having to hope bosses take their account of what happened seriously. Huffman says having a constant readout on productivity also helps managers be more responsive to nascent problems. "If somebody's struggling, not every associate is going to call for help," he says. "If we see their cycle time is jumping through the roof, we can go over and say 'Are you having any issues?'"Workers on Denso lines equipped with Drishti's technology now get a personal feed of their own data. Monitors on each workstation display how a worker is doing, says Raja Shembekar, a Denso vice president. If the worker completes their assembly step on time, they see a smiley face—if not, a frowny one. Hence, Amazon had begun to apply AI robotic to replace some warehouse workers' tasks.

For another factory manufacture working environment example, AI can replace many manufacture workers to do their tasks in factories. Route 9 skims by Boston and cuts clear across Massachusetts to Pittsfield, a city of roughly 50,000, the largest in Berkshire County. Well east of Pittsfield, Route 9 becomes Worcester Road, named for a city that in earlier times was the nation's largest manufacturer of wire—barbed wire, electrical wire, telephone wire and the wire used in the making of undergarments by the Royal Worcester Corset Co., once the largest employer of women in the United States. Older Worcester residents can still recall the factory bells pealing to signal the start and end of the workday. Now, the bells are silent, and the wire and corset factories have been replaced with three of the nation's largest employers: Walmart, Target and Home Depot. If this sounds familiar, it should. It has been nearly two decades since retail overtook manufacturing as the nation's most important job creator, employing roughly one of every 10 American workers—more people than in health care and construction combined. That's a lot of jobs.

Of course, not all retail jobs qualify as what most of us consider good jobs. Today, the average hourly wage for a nonsupervisory retail worker is $11.24, and less than half of retail workers receive benefits of any kind. Still, as a nation, we've come to a sort of uneasy peace with this trend. We know that manufacturing employs far fewer Americans today than it once did—that iPads and Macs aren't made in America and neither are many televisions, appliances, tools, toys or clothes. We also know that shopping for these appliances, tools, toys and clothes is an all-American pastime: On average, we spend nearly 45 minutes a day (more than 270 hours per year) purchasing goods and services. Retail has become the world as we know it, and many of us expect to make our living working in that world.Thanks to automation and a killer business model, Amazon is so efficient that it reaps nearly twice the revenue per employee of Walmart, despite the fact that Walmart, too, has a substantial online presence. Worldwide, Amazon has installed over 100,000 robots to labor in "perfect symbiosis" with humans in its warehouses and has plans to install many thousands more. While it's not clear what constitutes perfect symbiosis, the robots are said to save the company $22 million annually, per warehouse. The company's master plan of an autonomous future also includes goods delivered by drones and self-driving vehicles.

For while Amazon continues to open warehouses around the globe and staff them with many thousands of human beings, estimates are that every human on the Amazon payroll—whether full- or part-time—displaces two humans at traditional brick-and-mortar operations. And that's a feature, not a bug: As Tim Lindner, a veteran IT analyst, confided in a note to industry insiders, eradicating jobs is the explicit goal of any online retailer. As he once wrote: "Labor is the highest-cost factor in warehouse operations. It is no secret that Amazon is moving to highly automated operations within its distribution centers, and...it has additional technology that can further reduce the number of humans it needs to process customer orders.... You have heard the old programmer's phrase, 'Garbage in, garbage out.'... [With] the diminishing reading abilities of humans on the Receiving dock, finding an automated solution to eliminate the 'garbage in' problem is the holy grail. Amazon may have just patented it."

By garbage, Lindner meant human error, the alternative to which is apparently robotic precision. And robots can be very precise, especially when it comes to routine tasks. Sawyer, an industrial robot created by the former Boston-based Rethink Robotics, offers an impressive illustration of how all-embracing a robot arm can be. Sawyer is the brainchild of Rodney Brooks, the inventor of both Roomba, the robotic vacuum, and PackBot, the robot used to clear bunkers in Iraq and Afghanistan and at the World Trade Center after 9/11. Unlike Roomba and PackBot, Sawyer looks almost human—it has an animated flat-screen face and wheels where its legs should be. Simply grabbing and adjusting its monkey-like arm and guiding it through a series of motions "teaches" Sawyer whatever repeatable procedure one needs it to get done. The robot can sense and manipulate objects almost as quickly and as fluidly as a human and demands very little in return: While traditional industrial robots require costly engineers and programmers to write and debug their code, a high school dropout can learn to program Sawyer in less than five minutes. Brooks once estimated that, all told, Sawyer (and his older brother, the two-armed Baxter robot) would work for a "wage" equivalent of less than $4 an hour.

Robots loom large in discussions of work and its future, a conversation that can get mired in false assumptions. Until recently, many economists were skeptical that automation could permanently displace human workers on a large scale. People have always shifted away from work better done by machines, but the economic principle of "comparative advantage" predicts that humans will maintain an edge in many fields. Under this logic, technology will not displace us but set us free to do less dangerous, more challenging things, essentially the very things that make humans human. Of course, human workers are complicated. We get tired, hungry, distracted, angry, confused. We make mistakes, sometimes egregious ones. Machines lack our frailties and biases and are better equipped to weigh evidence fairly, without prejudice or false assumptions. Perhaps most critically, machines can retain and process data far more accurately than we can, and that data is growing exponentially.

Every minute of every day, Google services 3.6 million searches in the United States alone. Spammers send 100 million emails. Snapchatters send 527,000 photos, and the Weather Channel broadcasts 18 million forecasts. This and more data—properly collected, codified and analyzed—can be applied to automate almost any high-order task. Data can also serve as a surrogate for human experience and intuition. Online shopping and social media sites "learn" our preferences and use that information to make values-based assessments to influence our decisions and behavior. And, increasingly, machines excel in the tasks once thought uniquely human."Computers are able to see and hear, and have face-recognition capabilities that are significantly better than humans," says Vardi. "Machines understand the human world far better than they did just a few years ago. And we haven't discovered anything in the human brain that can't be modeled."

● AI can replace counter cashier service staffs

And robots need not be perfect, only equal to—or a tad better than—complicated and expensive humans. And technologists are working hard to make sure they are a tad better. For example, in the case of retail, it's become clear that many of us avoid the self-service checkout line—we prefer the cashier to punch in our purchases rather than do so ourselves. So it seems that the job of cashier—among the largest retail employment categories—is not directly at risk. But Zeynep Ton, an MIT management expert who focuses on the retail sector, says self-service checkout is only a first step and not a terribly smart one. "Customers recognized that self-service checkout is not an innovation, but merely a way of outsourcing the job to them, so they didn't like it," she says. "But new technology is coming that will make self-service checkout so much easier and faster, and that will have a real impact on retail employment."

Experts caution that the so-called apocalypse in retail

predicted a few years ago has not yet come to pass. In fact, for every company closing existing stores, two more are opening new stores. Retail is a highly competitive industry, and technology is transforming not only the way we shop but the way we connect with brands—for example, just a few years ago, who would have imagined that Amazon would open actual retail stores? And while e-commerce has grown to 10 percent of retail, that still leaves 90 percent for brick-and-mortar stores. But those brick-and-mortar stores, too, are undergoing radical change that has serious implications for America's workforce.

As example, Lobaugh cites food trucks, which he says increasingly pose a threat to many fast-food outlets. Unlike restaurants pinned down by a pair of Golden Arches, food trucks are nimble—they can home in on areas where customers are most likely to gather at any particular time. They can also tailor their offerings to a particular region or even a neighborhood, as well as use Facebook or other media to get out the word on their menu items and locations. Small, specialty stores also have far more flexibility than large department stores. "Technology has reduced the cost of entry into new markets, so in retail there are fewer big, monolithic companies, but more small competitors," he says. "Companies are diversifying to meet the specific needs and desires of consumers—everyone's piece is getting smaller, but there are many more pieces."

But despite what it predicts will be a banner holiday season, this year Amazon took on far fewer seasonal employees than usual—100,000 employees versus 120,000 the previous two years. And while an Amazon spokeswoman insisted that automation is not a factor in this reduced workforce, others seem to not agree. In a recent report, Morgan Stanley analyst Brian Nowak soothed the fears of

Amazon shareholders concerned with the wage increase by pointing out that automation had already and would continue to reduce the call for labor, and therefore reduce overall costs. When asked about this, Lobaugh again tactfully declined to comment—other than to say that while the retail sector had lost less ground than most people assume, retail employees were another matter. "There are winners," he says, "and then there are losers."

- AI can replace accountants in accountancy service industry

Not that long ago artificial intelligence (AI), robots and machine learning (ML) were thought to be things only found in science fiction films. Today, this type of technology is taking center stage in workplaces across the globe. Industries, including manufacturing, retail, agriculture, and customer service have already had AI replace some job positions that left workers scrambling to find new career options. This AI revolution is not expected to slow down anytime soon. In fact, experts anticipate that as many as 800 million jobs could be replaced with AI technology by the year 2030. Initially, AI technology and automation in the workplace seemed to only affect pink and blue-collar workers. As this technology advances and becomes more powerful, professional, white-collar workers, including accountants, are starting to worry about what the future holds for their career and if AI will be developed to own their professional skills in accounting service industry.

In basic terms, AI technology is intelligent machines that are able to complete repetitive, mundane tasks at a fraction of the time it takes humans and with greater accuracy. The emergence of Machine Learning now allows AI platforms to observe, analyze and self-learn data and processes to improve its performance and accuracy over time. AI technology is already able to handle many accounting functions, such as tax preparation, payroll, and audits. Many of the leading accounting software providers, including Xero, Intuit and Sage have incorporated AI technology into their software to handle basic accounting tasks, such as bank reconciliations, invoice categorization, risk assessment, and audit processes, like expense submissions and invoice payments. Many of these standard tasks are extremely time-consuming, which has many accountants across the country worried about how the emerging AI technology will affect their billable hours. An even bigger concern is that AI technologies will replace the need for companies to work with accountants at all.

- AI Will Transform not Replace Accountants

While there is no doubt that AI technology is capable of handling many standard accounting tasks faster and more efficiently or that these capabilities will only increase over time, it doesn't mean the end for accountants. There always will be a need for that human element - human intelligence - at the other end of AI technology. In fact, according to leading research firm, Gartner, AI is set to create more jobs than it will replace, leaving workers, including accountants with options. Accountants don't have to worry about their job being replaced by AI any time in the near future. Companies will always need accountants that can analyze and interpret AI data, as well as provide consulting services. Rather than replacing the role of an accountant, AI technology will transform the duties an accountant performs.

With AI technology and machine learning handling many of the mundane, repetitive tasks, accountants will have more time to focus on other aspects of the job, such as consulting and data analysis. This is good news for many accountants. Rather than spending hours completing menial tasks, accountants of the future will be able to use and analyze AI data to provide their clients with sound business solutions.

In many ways, AI will help accountants improve their services. AI technology will improve data entry accuracy and lower the liability risk for accountants. In addition, emerging technology is more efficient at fraud detection, adding an extra layer of protection for accountants and their clients. It also provides real-time data, which allows accountants to provide real-time solutions. Even more impressive is the ability of machine learning to analyze large amounts of data instantly, evaluate past successes and failures in an effort to accurately predict future outcomes.

There is no way to escape the use of AI technology, at least not if you hope to remain competitive in the upcoming years. The speed, efficiency and accuracy of AI technology just cannot be beat. The only thing accountants can do is to embrace this new technology and learn how to maximize its use. The better equipped you are to help your clients integrate and utilize AI technology in their accounting processes the more valuable you will be. For example, many universities today are already incorporating IT and database management courses into their accounting program. This means that graduating students are coming into the workforce with the skills they need for future accounting work. Accountants already in the workforce must find ways to acquire these skills in order to remain relevant to their employers and/or their clients. Accountants can obtain the IT skills they need by attending seminars, using self-learning online programs or attending college-level courses. It is equally important for accountants to stay up-to-date on the latest accounting trends, emerging technologies and industry news. This will allows accountants to not only keep their jobs but to also provide more efficient services to their clients. Rather than worry about AI taking over their jobs, accountants should embrace this technology as a powerful solution to enhance customer services. Finally, accountants will be able to use all their training and experience to provide customer will real and effective business solutions, whether it's in reference to tax consulting, real estate deals, mergers, growth options, or any other business practice.

On conclusion, technology is advancing at record rates so now is the time to obtain the IT and database management skills you need to advance into the future. With the right skills and training, accountants are guaranteed a lucrative career that will last well into the future.

5

(AI) -driven automation industry development

5.1 (AI) - driven automation industry development how to influence work nature change

(AI) -driven automation industry will create wealth and expand economy growth to any countries, but it will be accompanied by changed in the skills that workers need to learn. One of main ways that technology increases productivity is by decreasing the number of labor hours needed to create a unit of output. It implies (AI) technology will influence low educated and low skillful labor number to be decreased (reduction employment number).

In contrast, technological change tended to work in a different direction throughout the nowadays. The advance of computer and the internet raised the relative productivity of higher skilled workers. So, routine-intensive occupations that focused on predictable tasks disappearance, such as switch board, operators, filming checkers, travel agents and assembling line workers etc. were particularly replaced by new technologies.

However, today, it may be challenging to predict exactly which jobs will be most immediately affected by (AI) driven-automation. The reason is because (AI) is not a single technology, but rather a collection of technologies that are felt unevenly through the economy to influence job changing both negatively and positively. In positively view point, (AI) driven-automation will make many workers more productive and increase demand for certain skills. Consequently, new jobs are likely to be directly create in areas , such as the development and supervision of (AI) as well as indirectly created in a range of areas throughout the economy as higher incomes lead to expanded demand. Otherwise, in negatively view point, many traditional human needed (demand) skillful jobs will be threatened by automation are highly concentrated among lower-paid, lower-skilled and less -educated workers. It means automation will cause pressure on demand for this group, pressure and employment, if (AI) can replace the low skilled and less educated workers‘ jobs. Thus, (AI) will have negative influence to impact on the labor market.

(AI) capabilities will enable automation of some tasks that have long required human labor. Why can (AI) replace some simple human jobs? For example, advances in robotics are expanding machines' abilities to interact with and sharp the physical world. Combined , (AI) and robotics will give rise to smarter machines that can perform more sophisticated functions than ever before and brings more advantages that humans have exercised. This will permit automation of many tasks now performed by human workers and could change the shape of the labor market and human activity.

5.2 How (AI) influences labor market

Today, it may be challenging to predict exactly which jobs will be most immediately affected by (AI)-driven automation. Because (AI) is not a single technology, but rather a collection of technologies that are applied to specific tasks.

Some specific predictions are possible based on the current (AI) technology. For example, driving jobs and house cleaning jobs, bank counter service jobs, telephone enquiry service operators. Restaurant cooking jobs, simple accounting record service jobs etc. that require relatively less education to perform. Advancements in computer vision and related technologies have made the feasibility of fully appear more likely, potentially displacing some workers in driving-dominant professions. Seemingly similar robot, for which the operational tasks is less specific of navigating to a specific destination when following a set of given rules and preserving safety.

In the future, the effects of (AI) on the labor market in the decade ahead will continue the trend toward skill-biased change that computerization and communication innovations have driven in recent decades. Thus, some human driving occupation will be disappeared or replaced by (AI) automation driven. For example, bus drivers, light truck or delivery services drivers, heavy and tractor-trailer truck drivers, school drivers, tax drivers, travel bus drivers.

However, (AI) technology could enable some workers to focus time on other job responsibilities, boosting their productivity, and actually raised wage growth among those still holding the reshaped jobs. For example, salespeople, who currently spend a considerable amount of time driving could find themselves able to do other work when a car drives them from place to place, or inspectors and appraisers could fill out paperwork, when their car drives itself. This (AI) -driven technology should make these workers more productive, with (AI) -driven technology serving as a complement, not a substitute. New jobs will also likely be created, both in existing occupations cheaper transportation costs with lower prices and increase demand for products and all the related occupations, such as service and fulfillment, and in new occupations not currently foreseeable.

What kind of jobs will be created by (AI) technology? Predicting future job growth is extremely difficult, due to it depends on technologies or substitute for existing today as well as they may complement or substitute for existing human skills and jobs. However, (AI) will also lead to substantial indirect job creation to the degree it raises productivity and wages, it may also lead to higher consumption that would support additional jobs from high-end draft production to restaurant and retail. The future(AI) " augmented intelligence", the technology's role is as assisting and expanding the productivity of individuals rather than replacing human work. Thus, based on the biased-technical change framework, demand for labor will likely increase the most in the areas where humans complement (AI) automation technologies. For example, (AI) technology , such as IBM's Watson may improve early detection of some cancers or other illnesses, but a human healthcare professional is needed to work with patients to understand and translate patients' symptoms, inform patients of treatment options, and guide patients through treatment plans. Shipping companies may also partner workers who pick up and deliver products over the last feet with (AI) enabled autonomous vehicles that move workers efficiently from site to site. In such cases, (AI) augments what a human is able to do and allows individuals to either be move effective in their specially task or to operate on a larger scale. Thus, it seems (AI) technology will also create new jobs, raise productivities and workers' efficiencies.

Redefining management in
the workforce of artificial intelligence

- Change management

In the future, due to artificial intelligence influences to some kind of human jobs nature. So, the kind of human jobs of management methods will also need to change to adapt the artificial intelligence technology input to their organizations. It will cause challenges for every executive and manager if who won't have effort to manage their teams how to apply artificial intelligence technology to work efficiently and easily. For example, division of labor will change among humans and machines will increase. Thus, companies will have to adapt their training performance and talent strategies how to emphasize on work that how to make human judgment and skills and experimentation. Thus, (IA)'s greatest impact will be on administrative coordination and control tasks, such as scheduling , resource allocation.

In fact, mangers will encounter this challenges: How to apply human experience and expertise to judge critical business decisions and practices when the information available is insufficient to suggest a successful course of action? Due to this kind of work will require new skills and mindsets. I shall indicate these change management methods to adapt (AI) technology. Such as: administration and routine tasks, scheduling , allocation of resources and reporting will fall within the intelligence machines, responsibilities that have long been reserved for humans. For example, a typical store manager or a lead nurse at a nursing home most constantly arrange shift schedules, accounting for staff members' absences owing to illness, vacation time or sudden departures.

Thus, the managers need to learn how to arrange new division of labor within the organizations after (AI) technology had been implemented to the organization. Artificial intelligence is currently influencing into once considered exclusive to humans: assessing and acting on human emotions and personality traits. The influences to managers need to change their strategies to adapt (AI) technology implements include such as below:

Firstly, managers need to spend the bulk of their time on coordination and control tasks from intelligent system implements. Their time spending on these major three aspects from impact of intelligent system: coordinate and control, solve problems and collaborate and people and community , strategy and innovation three aspects. Thus (AI) will influence managers need to change their judgment method to teach whose teams how to adapt the (AI) system operations in any organizations.

Secondly, (AI) will influence top, middle and low level management needs to change to adapt the (AI) technology operations to any owned (AI) technology organizations in the future. Intelligent machines must be trained in context. Just like humans , on-the-job training is a requirement for such machines because they typically arrive with only very general capabilities. To get the most from (AI), managers at all levels must participate in the instructional experience and in the learning process and provides managers' familiarity with such systems on these aspects, e.g. How the system works and generate advice, how the system has a proven track record , how the system provides convincing explanations , how the system can make simple rule- based decisions.

Thirdly, managers need to learn how to make judgment more accurate (AI) systems assistance. Although (AI) will invariably take on more routine work and even augment human decision-making, it won't judgment work, the application of human experience and expertise to critical business decisions when the information available is insufficient to suggest a successful course of action or reliable enough to suggest an obvious course of action. For a sense of the nature of judgment work, consider big data marketing and sales analytics. Such analytics often provide insights that can inform promotional campaigns, including predicting which promotions will generate desired sales brand further into the future, marketing executives need use judgment, combining analytics with their own and others' insight and experience.

The application of experience and expertise to critical business decisions and practice represents the real value of human judgment. But, when artificial intelligent machines are implemented to any organizations to assist the low, middle and top level management to make any business judgment. These forms of judgment work that managers can gather data interpretation, idea development more absolute from (AI) machine assistance. Thus, why these level management executives need to learn how to apply (AI) machines to help them to make any business judgment more accurate.

- How (AI) influences organizational change

Consequently creative and social intelligence will be in even greater demand as (AI) makes in management and the workforce. This development will represent a long term trend in labor markets , one characterized by intensifying demand and reward for social skills with a growing desire for creative capabilities, managers will seek to fashion of ideas and hypotheses from inside and outside of the enterprise to shape solutions to their most pressing business problems. Thus, (AI) will influence overall organizational team members who have chance to participate any decision to make more accurate business judgment.

Many managers mistakenly view judgment work as only an individual discipline, failing to appreciate that it can also involve decide interpersonal and organizational practices. In more complex settings, judgment is typically a collective outcome of individuals' and teams' diverse perspectives, insights and experiences. And often , the resulting choices are better informed than decisions that an individual would have arrived at on his or her own.

Thus, when any organizations apply (AI) technology to assist managers to gather data and ideas to make any judgment. In these cases, organizations can create the conditions for effective collective judgment by establishing structures , such as " shadow advisory boards" that prompt managers and employees to source and synthesize multiple perspectives. Thus, a traditional organization (firm) might freshen its thinking is t put together a shadow advisory board, comprised of young, digital people who can apply (AI) machine assistance to make judgment work more accurate whether related to people development, problem-solving or strategizing and innovating for considerable degrees of creative and social intelligence.

Thus, on the one hand, (AI) technology machine augmentation and automation can give these advantages to human (organization managers) , e.g. developing people and community, solving problems and collaborating, coordinating and controlling work, shaping strategy and leading innovation. Besides, on the other hand, the next generation managers need have these individual attitude to treat intelligent machines to be as colleagues.

When, judgment is a human skill, intelligent machines can accelerate human learning that supports it, assisting in data -driven simulations, scenarios and search and discovery activities. Focuses on judgment work, some decisions require insight beyond what data can tell them. This is the sweet sport for human judgment, the application of experience and expertise to critical business decisions and practices. Thus, managers will also need to find ways to learn how to use digital (AI) technologies to tap into the knowledge and judgment of partners, customer external stakeholders and role models in other industries after the (AI) machine had been implemented to the organization.

5.3 Future works change: Automation, employment and productivity

● How (AI) influences employment

Human future " micro to macro" industry trends will be affected business strategy and public policy by (AI) technology. In the future (AI) technology will influence those six themes: productivity and growth, natural resources, labor markets, the evolution of global financial markets, the economic impact of technology and innovation and urbanization. However, (AI) technology will bring economic benefits of tackling gender inequality, a new global competition, Chinese innovation and digital globalization.

Nowadays, advances in robotics artificial intelligence, and machine learning are in a new age of automation, as machines match or outperform human performance in a development to any countries. For example, automation of activities can enable businesses to improve performance by reducing errors and improving quality and speed, and in some cases achieving outcomes that go beyond human capabilities. For example, some research indicated automation could raise productivity growth globally by 0.8 to 1.4 % annually; more than 2,000 work activities across 800 occupations. When less than 5% of all occupations can be automated using demonstrated technologies about 60% of all occupations have at least 30% of constituent activities that could be automated. Many occupations will change that will be automated away: Activities most susceptible to automation involve physical activities, in highly structured and predictable environments, as well as the collection and processing of data. They are most prevalent in manufacturing , accommodation and food service and retail trade and include some middle-skill jobs. For example, such as natural language processing is a key factor. Beyond technical feasibility, the cost of technology competition with labor including skills and supply and demand dynamics, performance benefits including and beyond labor cost savings, and social and regulatory acceptance will be affected by (AI) automation technology. Thus, (AI) automation will impact to influence global employment in those aspects as below:

Firstly, assuming that people are displaced by automation will find other employment. The anticipated shift in the activities in the labor force is of a similar order as the long-term shift away from agriculture and decreases in manufacturing share of employment. Both of manufacturing and agriculture industries which would be accompanied by the creation of new types of work not foreseen at the time.

Secondly, for business, the performance benefits of

automation are relatively clear. Thus, the businessmen have opportunities for their micro economies to benefits from the productivity growth potential and macro economies to benefit to encourage continued progress and innovation , investment and market incentives. At the same time, employers must innovate policies to help workers and institutions adapt to the impact on employment.

This will likely include rethinking education and training, income support and safety nets , as well as support for those dislocated, when employees need to leave themselves homes to move to other cities to learn new (AI) automation works. Thus, individuals in the workplace will need to engage move comprehensively with machines as part of their everyday activities, and acquire new skills that will be in demand in the new automation age. Consequently , the scale of shifts in the labor force over many decades that automation technologies can be a similar order to the long -term technology -enables shifts in the developed countries' workforces away from agriculture in the 21 th century. Those shifts did not result in long-term mass unemployment because they were accompanied by the creation of new types of work not foreseen at the time. However, human will still be needed in the workforce when the total productivity gains are caused by (AI) technology.

● What occupations will be influenced by (AI) technology.

In the future, scientists predict that these occupations will be influenced by (AI) technology mostly. They include : retail salespeople, food and beverage service workers, language or translation teachers, health practitioners. Since these work activities have a more relevant occupations are made up of a range of activities with different potential for (AI) automation . For example, a retail salesperson will spend more time interacting with customers, stocking shelves , or ringing up sales. Each of these activities is distinct and requires different capabilities to perform successfully.

Thus, these job activities have similar simple control characteristics. Simple activities include greet customers, answer questions about products and services, clean and maintain work areas, demonstrate product feature process sales and transactions. All these activities can have similar simple activities in order to (AI) machines can be learn how to do these activities from (AI) technology . For example, the capability perception includes sensory perception, cognitive capabilities, such as retrieving automation, recognizing known patterns(supervised learning), logical reasoning problem solving.

Thus, (AI) machine is such human, which has feeling and emotion, such as social and emotional sensing, judgement reasoning methods, natural language understanding and physical capabilities, such as mobility , navigation, gross motor skill, fine motor skills. It seems that the future, (AI) human invents machines which will have these human characteristics to do human similar behavioral job duties more easily and efficiently. It implies these above human occupations will be replaced by (AI) human invention machines in the future. Due to (AI) creation, it is possible to cause unemployment number of these above workers will increase because (AI) machines can do their similar job behavioral activities.

Consequently, employers won't need to employ many of these skillful labor. Otherwise, they can buy less number (AI) machines to attempt to do whose job activities more easily and efficiently. So, it seems (AI) machines will have more high work performance to replace these occupation workers' work performance. Finally, these occupation worker unemployment number will only increase when the (AI) machines had been invented to achieve to do their work behavioral activities absolutely success in the future.

- Whether (A) technology machine labor will replace human worker more or assist human worker more

There is no single agreed definition of a robot how outcome of a task that is completed without human intervention. When some definitions require the task to be completed by a physical machine moves and respond to its environment, other definitions use the term robot in connection with tasks completed by software , without physical embodiment.
However, to answer the question : Whether (AI) technology machine labor will replace human worker more or assist human worker more. I shall indicate some examples to let readers to judge whether (AI) technology can create new jobs or reduce old jobs.
Firstly, I shall explain what (AI) function is. (AI) is a service robot that performs useful tasks for humans or equipment excluding industrial automation application . Thus, the classification of a robot into industrial robot or service robot is done according to its intended application. It is also a personal service robot or a service robot for personal used for a non commercial task, usually by lay persons . Examples are domestic servant robot, and pet exercising robot. It is also a professional service robot or a service robot for professional used for a commercial task, usually operated by a properly trained operator. Examples, are cleaning robot for public places, delivery robot in offices

or hospitals, fire-fighting robot, rehabilitation robot and surgery robot in hospitals. Thus, these functions will be future (AI) application to our daily life necessaries or business necessaries.

However, some authors agree (AI) will bring negative outcomes of automation, due to raise competiveness, reduce human job nature. Otherwise, other authors argue (AI) will bring positive outcomes of automation, due to raise productivities, job creation, assist humans work.

On the positive outcome hand, robots can increase productivity . This is particularly important for small-to medium sized businesses both are in developed and developing countries economies. It also enables large companies to increase their competitiveness through faster product development and delivery. Increased use of robot is also enabling companies in high cost countries to re shore, or bring back to their domestic base parts of the supply chain that will have previously outsourced to sources of cheaper labor. Currently , the greater threat to employment is not a automation, but an inability to remain competitive. Automation has led overall to an increase in labor demand and positive impact on wages. The reason is that the middle-income/middle-skilled jobs have reduced as a proportion of overall contribution to employment and earnings leading to fears of increasing income inequality, the skills range within the middle income bracket is large. Thus, robots are driving an increase in demand for workers at the higher -skilled and with a positive impact on wages. This issue is how to enable middle-income earners in the lower-income range to unskilled or retain. Finally, the (AI) positive impact supporter who argue the future will be robots and humans can work together.

However, on the negative outcome hand, robots can

substitute labor activities, but don't replace jobs. They believe that less than 10% of jobs are fully automatable. Increasingly , robots are used to complement and augment labor activities, the net impact on jobs and the quality of work is positive. Automation can provide the opportunity for humans to focus on higher-skilled, higher-quality and higher-paid tasks. Robots can improve productivity when they are applied to tasks that which perform more efficiently and to a higher and more consistent level of quality than humans. For example, increased productivity is enabling some firms, such as Whirlpool, Caterpillar and Ford Motors company in the US restructure their supply chains, bringing back parts of the manufacturing process to the country of origin. Thus, productivity gains due to robotics and automation are important not just at the company level, but also for build industry and nation competitiveness.

I suppose that productivity can be raised. What are the impacts of robots on employment? Firstly, the main focus of development has been on personal entertainment, which does not drive worker productivity (manufacturing production). When the internet (information and communication technology (ICT)) innovation. This is borne and by findings that manufacturing productivity, which has been driven by innovations in automation rather than consumer technologies, has government strongly than productivity in the services sectors of the economy in most nature economies. It seems (AI) automation will create many jobs in internet communication entertainment game industry. For example, many young people like to use internet to play any electronic games from computer or mobile at home or outside home conveniently. Thus, (AI) automation will increase demand to be invented to any new entertainment game from internet channel. It will need to employ many (AI) entertainment game inventors to create many automation entertainment games. Thus, (AI) automation in internet entertainment game industry will need human (AI) entertainment game inventors to invent the knowledge-based capital of (AI) automation entertainment games. The (AI) entertainment game inventors will need own research and development skills, form specific skills, organizational know-how skills, databased knowledge, design and various forms of intellectual property to do these (AI) automation entertainment game invention occupations in the future.

International Federation Of Robotics(2016) indicated that China will be as a major robotics manufacturer and user of robots, benefiting from jobs created by robot manufacturing and productivity gains from robot use. Chins had sold of robots to any one single market every year since 2017 year. The Chinese government has included a focus on robotics in its 10 year strategy. In order to achieve its target of a robot density of 150 units per 10, 000 workers by 2020 year. Thus, Chinese companies will have to install around 650,000 new industrial robots between 2016 to 2020 year, 2.5 times more than installed globally in 2015 year.

Hence, China (AI) manufacturing industry will need to employ many workers . It implies (AI) manufacturing industry will create many new occupations in China. Also, ministry of economy, trade and industry (2015) also showed that Japan currently has the largest stock of industrial robots in operations, primarily in the automation industry. Driven by a rapidly aging population and low productivity rates, the Japanese government has sights on a 20-fold increase in the use of robots in the non-manufacturing sector and a three-fold growth rate of labor productivity in the service sector both by 2020 year. Thus, it also implies Japan will need many robots to be provide to service industry. Due to robots will provide to serve any businessmen's clients. Thus, it is possible that the service workers won't be dismissed as well as it is depended on the serving job nature to decide whether Japan's service workers can still serve to their employer when the service (AI) robots are applied to whose employers.

Consequently, it seems that (AI) can create employment, Ministry of economy, trade and industry (2015) showed that such as China will develop the major (AI) automation manufacturing industry. The (AI) employers will need to employ many workers to manufacture any these different kinds of (AI) robots to satisfy China or overseas individual or business buyers needs. But, (AI) can also cause unemployment to the low skillful service workers. Such as if Japan some service businesses choose to buy any (AI) service robots to replace their service staffs to serve their clients. It is possible that the service staffs will be dismissed, due to (AI) robots can do such as their same service job duties to achieve better service performance.

Thus, today, it is increasingly common for people to use robots in various situations at home and in retail stores, hotels and hospitals these service industries. Robots are classified into server types based on their functionality (service and utility robots or those designed to communicate with humans) and appearance (humanoid robots or mechanical robots). The type of robot, to which each country allocated particular importance in the advance of robotics, reflects the sense of values and preferences of its population. Thus, if the country has high population needs to use robots, then they will influence either more new jobs creation or more old job loss in the country's (AI) manufacturing or (AI) service industries both. For example, Japan respondents often associate the term " robot " with humanoid robots that can communicate with human and they have a high level of familiarity with robot. The US has the highest level of robot utilization at home and in retail stores with its people being the most enthusiastic about the future use of robots. Germany shows a strong tendency to consider robots for industrial purposes and its people feel strong effort to the presence of robots in their households.

In conclusion, to judge whether how (AI) will influence the country's employment to be better or worse. It will depend on the country home buyers (users) or business buyers (users) how to use (AI) for their daily needs. If the country , such as US retail stores need to use (AI) , it will have possible to reduce some or many retail service workers. Even, if the country , such as Japan has many home users need to use (AI) , it will not influence the employment market. Otherwise, it will raise (AI) salespeople numbers. Even, if the country, such as Germany and China will have many (AI) manufacturers, then it will create many (AI) manufacturing occupations for these (AI) manufactory workers.

Consequently, (AI) robots manufacturing and service needs will have positive or negative impact to any country's employment. It will depend on the (AI) service provision and service workers' job nature as well as the manufacturing workers of (AI) knowledge level to decide their employment chance in their country's employment market.

5.4 How can (AI) influence labor market?

- How can human society job nature
to be changed to artificial intelligent society?

From the first intelligent perspective reason view point, artificial intelligence is making machines " intelligent" acting as humans expect people to act. Artificial intelligence has ability to distinguish computer responses from human responses, it owns knowledge to solve expert problem. From another research perspective reason view point, artificial intelligence is the study of how to make computers do things which, at the moment, people do better (Rich & Knight, 1991, p.3).

(AI) researchers are native in a variety of domains, e.g. formal tasks (mathematics, games), tasks (perception, robotics, natural language, common sense reasoning), expert tasks (financial analysis, medical diagnostics, engineering, scientific analysis and other areas).

From the second business perspective reason view point, (AI) is a set of many powerful tools, and methodologies for using those tools to solve business problems. From a programming perspective reason view point, (AI) includes the study of symbolic programming problem solving and search .

From the third human technological perspective reason view point, today's computer can do many well-defined tasks, for example, arithmetic operations, are much faster and more accurate than human beings. However, the computers‘ interaction with their environment is not very sophisticated yet. How can human test whether a computer has reached the general intelligence level of a human being? Can a computer convince a human interrogator that it is a human? But before thinking of such advanced kinds of machines, human will start developing our own extremely simple " intelligent" machines.

So, it is possible that human society job nature will to be changed to artificial intelligent society when (AI) technology is developed to the mature stage in the future.

- Why does human need artificial intelligence machines?

One of major division in (AI) is between humans who think (AI) is the only serious way of finding out how we (human) work and human who want companies to do very smart things, independently of how we (human) work. This is the important distinction between cognitive scientists vs engineers. One of another major division in (AI) is between symbolic (AI), which represents information through symbols and their relationships. Specific Algorithms are used to process these symbols to solve problems or deduce new knowledge and connectionist. So (AI) , which represents information in network. Biological processes underlying learning, task performance and problem solving are imitated from human mind behaviors.

Thus, it is possible that artificial intelligence machines can do the better judgicious behavior to compare human.

- How does artificial intelligence influence future working changing in automation employment and productivity aspects?

In the automation changing influence aspect, as companies increasingly use robots on production lines or algorithms to optimize their logistics manage inventory, any carry out other core business functions. Technological advances are creating a new automation age in which ever-smarter and more flexible machines will be deployed on an ever larger scale in the marketplace. However, researching artificial intelligence with how influences human working nature. We need to answer these questions: How will automation

transform the workplace? What will the implications for employment? And what is likely to be its impact both on productivity in the global economy and on employment?
Advances in robotics, artificial intelligence, and machine learning are growing in a new age of automation as machines match or outperform human performance in a range of work activities, including ones requiring cognitive capabilities. What factors are determined the changing in workplace adoption by artificial intelligence innovation? What advantages are automation? Automation of activities can be enabled businesses to improve performance by reducing errors and improving quality and speed, and achieving outcomes that go beyond human capabilities.
Some scientists indicated based on their scenario modeling. They estimated automation could raise producing growth globally by 0.8 to 1.4 percent annually. Almost, the activities people are paid almost $16 trillion in wages to do in global economy have the potential to be automated by adopting currently demonstrated technology. According to their analysis of more than 2,000 work activities across 800 occupations. When less than 5% of all occupations have of least 30% of activities that could be automated. They also indicated that technical economic and social factors will determine automation. Continued technical progress, for example, in areas such as natural language processing is a key factor beyond technical feasibility , the cost of technology, competition with labor including skills, and supply and demand dynamics, performance benefits including and beyond labor cost savings and social and regulatory acceptance will affect (alter) the scope of automation.
Other some scientists also indicate U.S. country for example, the anticipate shift in the activities in labor force

of a similar order of magnitude as the long term sight away from agriculture and decreases in manufacturing. Share of employment in the United States both which were achieved. So, those factors can influence why artificial intelligence technology needs. So, it is possible that future agriculture and manufacturing both industries will apply (AI) technology manufacturer-kind of job nature to raise productivity instead of farmers, fruit picking workers, farming transportation labours as well as factory manufacturing workers and supervisors etc. human-kind of job nature.

- Is artificial intelligence possible to replace labor ?

Not just intelligence, but also debating, if machines are capable of having a conscious minds. Artificial intelligence has those characteristics as below:

On functionalism aspect, artificial intelligence inputs mental states, sensory inputs, (beliefs, desires being in pain feeling) and behavioral outputs. Since mental states are identified by a functional role, which are thoughts to be manifested in various systems. Even, perhaps computers which are physical devices with electronic substrate that inform computations on inputs to give outputs similar to brains which are artificial intelligence composed of part any intrinsic relationship to each other. Thus, artificial intelligence activities is not the whole itself, but into parts or on external influence on the parts.

On dualism aspect, artificial intelligence is a set of views about the relationship between mind are matter. On materialism aspect, it builds the only thing that exists is matter, including consciousness.

On biological naturalism aspect, it is similar a human brain than feels pains makes mental situation. So, artificial intelligence is similar biologist which might to be excited to human labor work. Hence, it seems artificial intelligence can change (alter) or replace human labor work of nature in possible in the future.

- Can (AI) technology replace human labour nature of work?

On technological innovation reason view point, the history development of artificial intelligence studying the intelligence is one of most ancient scientific discipline. The history development of artificial intelligence what aims to achieve human use to sense, learn remember and think, logic probability, decision making and calculation develop from mathematics, instead of replacement human labor functions.

Artificial intelligence history development aim is the scientific analysis of skills in connection and practice with the appearance of computers from 1950 year beginning. The artificial intelligence (AI) can deal with the ultimate challenges. How can (either biological or electronic) mind sense, understand and manipulate a world that is much simple and more complex than itself? And what if would human like to construct something with such capabilities?

The general-purpose software of the early period of (AI) were only able to solve simple tasks effectively and failed when which should be used in a wider range or an more difficult tasks. One of the sources of difficulty was that early software had very few or mix knowledge about the problems which handled, and activities successes by simply syntactic manipulation. Moreover, the other difficulty was that many problems that were tried to solve by the (AI) were untreatable.

The early (AI) software whether trying step sequences based on the basic facts about the problem that should be solved, experimented with different combinations till which found a solution. From the end the 1960 year, developing the so-called expert systems were emphasized. These systems had (sue-based) knowledge base about the field which handled. Till to the beginning of the 1970 year, (Prolog) the logical programming language was born, which was built in the computation realization of a version of the resolution calculus. (Prolog) is a remarkably prevalent tool in developing expert systems (on medical, judiciary and other scopes), but natural language parsers were implemented in this language. Then, in 1981 s, the Japanese announced the fifth generation computer system project, a 10 years plan to build an intelligent computer system that use the (Prolog) language as a machine code. Nowadays, (AI) can be applied any industries, such as car manufacturing industry can use (AI) technological machine-men manufacture car, instead of replacing human labors in factory. Even, in the future, using (AI) machine-men drivers can drive any private cars or public transportation tools, instead of replacing human drivers, e.g. bus, train, tram, ferry etc. Also in the future, machine-men can replace housewives to serve families to do housekeeping clean job , e.g. cleaning toilets, bathrooms, kitchens, even cooking functions at home. So (AI) machine-man can reduce housewives works at home. Moreover, (AI) machine man can take care old people , when who are living at homes or elder care centers.

So, it seems artificial intelligence (AI) will be possible developed to manufacture a new generation machine-man to assist (serve) families to do any simply cleaning or cooking jobs at homes. Moreover, the overall demand of (AI) general social needs will also rise, such as security, driving transportation tools, restaurant cleaning, elder centers care service etc. So, it seems that individual or families or social needs of (AI) will be increase in the future. Thus, it will influence macro economy growth (GDP) if there are large house family consumer group and hotel or bus or taxis or ferry etc. different business consumer group demand any artificial intelligence machine numbers increasing. Then, the artificial intelligence products and material manufacturers must need to buy many artificaial intelligence materials to produce any kinds of artificial intelligence machines to prepare to satisfy consumer individual needs. Consequently, macro economy will grow to the owned artificial intelligence development countries, e.g. US, China, UK.

- Why can artificial intelligence satisfy human needs?

First, On machine-man satisfactory demand aspect view point, it makes computers that think, it is the automation of activities. We associate with human thinking: like decision making, learning. It is the act of creating machine that perform function that require intelligence when performed by people. It is the study of mental faculties through the use of computational models. It is the study of computations that make it possible to perceive, reason and act. It is a branch of computer science that is concerned with the automation of intelligent behavior. It is anything in computing service that human don't yet know how to do property.

Second, on thought aspect artificial intelligence means systems thank think like humans, systems that think rationally.

Third, on behavioral aspect, artificial intelligence systems that act like human and that systems act rationally. However, the basic objective of (AI) is to represent human's thought processes in computation . These machines are supposed to exhibit behavior that. It is performed by a

human being, would be considered intelligent. However, some authors feel (AI) has disadvantages, such as it is not creative, it is excited in the use of sensory devices, it can't make use of a very wide context of experiences and it does not use common sense.
For speech recognition and understanding function needs example, (AI) can be applied in speech recognition and understanding function, which (AI) speech or voice recognition is a data input method. For example, the computer recognizes and understands one (or a few) word commands. Speech understanding on the other hand is the computer's ability to understanding a spoken language. That is , the computer understands the meaning of sentences, an paragraphs through (AI).
So, (AI) can be attempted to learn human language how to speak. It is similar to translate human language skill, instead of actual human speaking skill. Also, (AI) can assist handicap learning or language student how to listen different languages by machine-man sounds from computers more accurately. So, it seems that it (AI) can replace human language teachers speaking function and can change teaching language nature of job in language speaking and listening education industry.

- Is artificial intelligence one good choice for human future technological benefit?

Nowadays, new technology development is popular. However, artificial intelligence is one kind of new technology choice among different technologies innovation. So it brings this question: Is artificial intelligence technology value to invest? To answer this question. I shall indicate some other new technology developments to compare (AI) technology development to judge which has urgent needs to achieve human expectation nowadays.

For example, why is green peace interested in new technologies? New technologies features prominently in our ongoing campaigns against genetic modified crops and number power. However, which are also an integral part of our solutions to environmental challenges, including renewable energy technologies, such as solar, wind and wave (water) power energy as well as waste treatment technologies, such as mechanical, biological treatment. It seems humans need concern how to apply (AI) technology to solve environment pollution challenges in our future. So, environment protective, agriculture, natural energy technology will be popular demand to attempt to apply (AI) technology to solve their challenges or apply (AI) to assist to develop their industry.

- How can artificial intelligence impact on workplace?

Modern information technologies and the labor economy growth of machines is powered by artificial intelligence have already strongly influenced the world of work in the 21 ST century. Computers, algorithms and software simplify every tasks and it is impossible to image how most of our life could be managed without them. How can be the information economy characterized by exponential growth replaces the most production industry based on economy of scales? What will the future world of work look like and how long will it take to get? Will the future world of work be a world where humans spend less time earning their livelihood? Alternatively, are mass unemployment, mass poverty and social distortions also possible scenario for the future, where robots, artificial intelligence systems play an increasingly central role? These questions concern how artificial intelligence further development . Can influence labor economy growth on workplace ? When the labor market has widespread impact on intelligence property, information technology, product liability, competition and labor and employment laws.

How (AI) technology impacts on labor workplace.

The future influence any organizations how labor economies use of (AI) can be analyzed, such as deep machine learning is based on a set of model high level data. Unlike human workers, the machines are connected the whole time in workplace. If one machine makes a mistake, all autonomous systems will keep this in mind and will avoid the same mistake the next time.

Over the long run intelligent machines will win against every human expert. Production robots have been replacing employees because of the (AI) technology. They work more precisely than humans and cost loss. Creative solutions like 3D printers and the self learning ability of these production robots will replace human workers, the automatic data recording and data processing, traditional back office activities are no longer in demand. Autonomous software will collect necessary information and will send it to the employee who needs it. Additionally, dematerialization leads to the phenomenon that traditional physical products are becoming software. For example, CD or DVDs are being replaced by streaming services. The replacement of traditional event ticket, e-travel ticket service products or hard cash will be the next step, due to the possibility of payment by smartphone. So, (AI) technology will impact human's daily life consumption behaviors in the future. For another example, transportation tools, such as boats and ferries and private vehicles will use sensors and navigating without human input. Taxi and truck drivers will become obsolete, the stock store applies to stock managers and postal carriers of the delivery is distributed by (AI) machine delivery method.

What is the relationship between (AI) and (CRM)?

- Can (AI) technology impact on customer relationship management (CRM) ?

Nowadays , (AI) is a technology almost as old as the computer industry itself, it is similar with the advent of personal assistants function to businesses and personal promotion channel, such as (Amazon's Alexa, Apple's Siri, Google's Assistant) image recognition (face book), personalized recommendations (Netflix , Amazon). Those innovations have been driven by a increase in processing power, lower cost hardware, and the exploding creation and availability of data. It seems, (AI) technology can impact global customer service management method.

How to forecast economic impact modeling to (AI) will affect global economy? Can human forecast business revenue growth and job creation (or destruction) based on (AI) applied to customer relationship management (CRM) activities? In addition to the economic impact on (AI) or (CRM) which can include an estimate of the economic impact attributable to sales forces customer base. What can economic benefits be brought to (CRM) from (AI) technology?

Artificial intelligence(AI) comprises a set of technologies that use natural language processing, machine learning, knowledge graphs, and other tools to answer questions, discover insights and provide recommendations. Computer systems can use (AI) hypothesize and formulate possible answers based on available evidence can be trained through the ingestion of vast amounts of content, and automatically adapt and learn from (AI) self mistakes and failures.

So, any business organizations (customer service departments) can provide efficient and effective customer relationship management of excellent customer service quality if which applied (AI) technology system. The different type of (AI) systems include: (AI) system platforms, machine learning (AI) based data preparation and enrichment tools, machine vision/image recognition, voice speech recognition, text analysis and natural language processing, bots , e.g. face book website and virtual digital assistance solutions, social media pattern analysis , sentiment analysis, advanced numerical analysis (e.g. IOT streaming , machine logs), supporting technologies, knowledge base dialog management, Q&A processing etc. different (AI) technology system customer relationship management (CRM) tools.

(AI) (CRM) of activity can include these categories, such as: corporate marketing, marketing operation, field marketing, customer support, digital commerce, customer analytics, customer influenced product or service design, product or service pricing, finance information, presentation, customer billing, inventory , logistics and fulfilment support, partner management etc. different CRM tools.

(AI) technology of CRM has been carrying on plan different stages to achieve CRM personal assistant tool for businesses. The stages are such as, in the beginning stage of (AI) projects in place, implement now, pilot phase next year in the final stage of (AI) customer relationship management tools are foreseeable future. So, this CRM technology has been improved to plan in different stages every year to prepare to achieve full capacity of CRM service quality for businesses to use in the future.

Hence, how to develop an estimate prediction of the economic impact (AI) technologies could have CRM activities, which depends on gathering macroeconomic information on business revenue and the basic marketing of business revenue and the basic markup of business expenses by major functions (customer support, marketing and sales , production etc.)

An economic impact model that can gather data together and forecast the results how (AI) artificial intelligence technology brings (CRM) customer relationship management benefits to businesses, e.g. surveys investigation includes IT spending by sample countries, GDP and population estimates and forecasts, revenue per employee and ratios of IT spend to GDP. Surveys (questionnaire questions) of forecast results are influenced by (AI) impact can include: results are projected from surveys and rely on estimates are made by respondents on the expected financial improvements in categories of (AI) –assisted customer relationship management activities. The forecast assumes that these estimates are correct; financial estimates are based on estimates of "first year" improvement from full (AI) implementation; forecasts are from planning to implement any artificial intelligence of customer relationship management (CRM) projects, the improvement

forecast is of categories of activity , e.g. corporate marketing , digital commerce, and customer analytics. They are not estimates of ROI for the (AI) software. They rely on conservative estimates to which each of these entities might affect company revenue, expenses or productivity. They also rely on estimates of the penetration of software in customer relationship management activities . Net new jobs created are based on the ratio of new revenue to jobs required to support that revenue . They can assume that 50% of the net new revenue will support increases in labor and the rest will go for capital and other operating expenses that may replace jobs lost to automation.

In the future, some of the ways in micro economic benefits to any organizations. (AI) technology is expected to impact CRM activities include: Spending up sales cycles, improving lead generation and qualification solving customer support problems faster (raising service quality), helping companies improve brand campaigns and recognition, lowering costs of support calls when increasing resolution rates, lowering the cost of recruiting employees and partners, increasing revenue from optimized product marketing, optimizing price, distribution logistics and preventing loss through fraud detection. So, micro economic benefits view point, it seems that (AI) CRM technology can raise any companies economic benefits for care term.

Artificial intelligence enables machines or the in-build software to behave like human beings which allows these decisions and act. The advent of (AI) is leading , talking, making decisions and act. The advent of (AI) is leading to new technologies advances and transforming the economic and employment opportunities for humans in a positive way. (AI) related technologies can facilitate our live. For example, industrial robotics, robotic medical assistants, smart games, financial forecasting software, big data analysis, algorithms in health and bioinformatics, pilotless cargo places, drone ambulances and general purpose and workplace robots and others. (Disruptors technologies: Advances that will transform life, business and the global economy).

Artificial intelligence also known as computational

intelligence is defined as " the human –like intelligence exhibited by machines or software. It is theorized that intelligence of humans can be described and intelligence machines or software can simulate it. These machines software can be reasonable , learn, perceive and process information, like human mind and thus facilitate human life. They can think and act for us. So, artificial intelligence is an interdisciplinary field of study including computer science, neuroscience, psychology, linguistics and philosophy.

However, (AI) research and developments have economically impacted many industries, such as robotics, telecommunications, computer applications , health, finance, heavy manufacturing, transportation, aviation, e-service and e-commerce, military , music and movie, toys and games entertainment etc. industries.

In fact, many ideas, systems and technologies have been developing in the world of (AI) technology. However, which are net called or considered (AI) products, rather which are mentioned with their specific names, such as smart graphics, machine learning, e-commerce etc. (i.e. this is called (AI) effect).

- How does (AI) technology influence

the future of employment change?

Are future nature of jobs changed to computerization from (AI) technology? Where are the probability of computing occupations from (AI) technology influence? What is expected impacts of future computing on labor market from (AI) technology influence? John Maynard Keynes's frequently cited prediction of widespread technological unemployment " du to our discovery of means of economic the use of labor outrunning the pace of which we can find new used of labor" (Keynes, 1933, p.3).

In the future, (AI) technology will impact some nature of occupations to change computing. This chance will also influence some countries' economic change. For example, some factory human labors hand routine manufacturing tasks will be changed to computerization of routine manufacturing tasks by (AI) technological machine men hand manufacturing method. it will cause a structured shift in the labor market, with workers reallocating their labor supply from middle-income manufacturing to low-income service occupations.

Arguably, this is because the manual tasks of service occupations are less computerization, as who require a higher degree of flexibility and physical adaptability. So, (AI) technology will influence the human hand labor skillful occupation nature of task cheaper , such as vehicle manufacturing , ship manufacturing, computer manufacturing, steel manufacturing, television, radio etc. home electronic products of heavy machine industry change. Due to (AI) technology machine man will be proper to be used to manufacturing these electronic products when the (AI) technology innovation can develop to the mature stage. Then, any countries manufacturers will choose to use (AI) technology machine man, instead of human hand production.

Supposing the future prices of computing are fallen, seriously, problem solving skills are becoming relatively productive, explaining the substantial employment growth in manufacturing occupations, involving cognitive tasks where skilled labor has a comparative advantage, as well as the increase education needs for (AI) technology computing of machine man subject study.

Prediction of education needs for (AI) technology student numbers will increase, due to manufacturing industry needs many (AI) technology students in future employment market. Another (AI) technology influence if the future (AI) technological innovation, e.g. machine man manufacturing or machine man service industries will both increase demand, then with more sophistic software technologies will be disrupted labor markets by marketing workers redundant.

For publishing industry, what is striking about the case in paper book publishing industry will be unpopular? Due to the electronic book publishing industry will be popular, e.g. Amazon publish . (AI) technology can influence paper book manufacturing method which is replaced by machine man electronic book manufacturing method as well as it will cause the computerization is no longer confined to routine manufacturing tasks. Due to (AI) machine man manufacturing technology will be proper to be used to manufacture any products in short time efficiently and effectively , e.g. electronic book products. In the future, if it is fact to occur this case, such as (AI) technological machine man manufacturing method will be adopted (applied) to manufacture electronic books or any products in possible. (AI) technology will cause many manufacturing workers are unemployed. It is beneficial to employers, who can reduce to spend much wages expenditure to employ manufacturing workers, but it will cause many manufacturing workers loss jobs and reduce income to support whose families lives. It will cause social challenges, e.g. increasing stealing crimes if the manufacturing workers had not other skills to find other jobs to do easily. So, manufacturers need to concern over technological unemployment which will be hardly future phenomenon if who decided to dismiss all manufacturing workers, due to (AI) technology machine men replace to them.

If (AI) technology can be innovated to produce any kinds of machine man to serve any service or manufacturing industries successfully. Then, it will bring these questions: Can future that workers be influenced to be automation employment and productivity by (AI) technology influence? Does it impact to influence the (AI) technology countries' productivity and growth and natural resources development and labor markets and evolution of global financial markets and economic impact of technology and innovation and urbanization etc. issues? How will automation transform the workplace? What will be the implication for employment? What is likely to be its impact both on productivity in the global economy and on employment?

In fact, automatic of activities can enable businesses to improve performance by reducing errors chance and improving quality and speed, and same cases achieving outcomes that go beyond human capabilities. Some economists indicate (AI) technology would give a needed boost to economic growth and prosperity have of the working age population in many countries. Based on the scenario modeling, they estimate automation could raise productivity growth globally by 0.8 to 1.4 % annually. They also indicated that almost half the activities people are almost $1.6 trillion in wages to do in the global economy have the potential to be automated adapting current demonstrates technology, according to their analysis of more than 2,000 work activities across 800 occupations. When less than 5% of all occupations can be automated entirely using demonstrated technology, about 60% of all occupations have at least 30% of worker made activities, that would be automated. More occupation will change to be automated. They also indicated for business performance benefits of automation are relatively clear, but the issues are more complicated by policy making to attract foreign investors. Beyond technical feasibility, the cost of technology, competition labor will include skills and supply and demand dynamics, performance benefits and beyond labor cost savings and social and regulatory acceptance will affect the automation. Their predictions suggest that half of today work activities could be automated by 2055 year, but this could happen 10 to 20 years earlier or latter depending on the various factors in addition to their wider economic condition.

Some scientists suggest (AI) technology is finally starting to deliver real-life business benefits. Computer power is growing significantly , algorithms are becoming more sophisticated and perhaps most important of all, the world is generating vast quantities of the fuel that powers (AI) technology data billions of gigabytes of it every day. Also, online firms are digital natives, such as Google online search service company is investing on (AI) technology. For new though most of the news if coming from the suppliers of (AI) technologies. And many new users are only in the experimental phase. Few products are on the market or are likely to arrive these soon to drive immediate and widespread adoption. As a result, analysts believe (AI) technology's potential will give true economic benefit in the future. (AI) industry will introduce to suppliers and users to raise economic potential of (AI) technology.

In the future, (AI) technology systems can solve business problems. Some scientists categorized those into five technology systems that are key areas of (AI) technology development: robotics and autonomous vehicles, computer vision language virtual agents and machine learning , which is based on algorithms that learn from data without

replying on rules-based programming in order to draw conclusions or direct an action.
Such as computer vision and language includes natural language processing, analytics, speech recognition technology, some are about learning from information, such as about machine learning and others are related to acting on information, such as robotics, autonomous vehicles and virtual agents, which are computer programs that can converse with humans. Machine learning and a subfield called deep learning are artificial intelligence applications.

- Can artificial intelligence impact global office productive efficiency ?

Artificial intelligence (AI) is a term first defined in 1956 year. It is a branch of computer science that aims to create intelligent machines that work and react like humans. In contrast today, 60 years later, (AI) is characterized by a number of applications, including computers playing games against humans and understanding human languages, virtual personal assistants, and robotics which involve computers seeing , hearing and reacting to sensory stimuli. In the future, technologists predict for (AI) technology ranging from (AI) being used as a tool to aid relatively simple processes for robots with human like mental capabilities, who expect (AI) technology can emulate human performance by learning, coming to mind its own conclusions, understanding complex content, engaging in dialog with people, enhancing human cognitive performance or replacing humans in executing both routine and non-routine tasks. In existing industry, (AI) technology is used , such as targeted advertising and virtual used personal assistant as well as the (AI) technology that my exist in the future, such as robots with human vehicle processing capabilities.

The range of (AI) technology's progress in the future will determine the economic impact future of (AI) technology on the global economy with more limited advances and applications (i.e. weak (AI) only) corresponding to more limited economic impacts and more substantial progress, i.e. strong (AI) technology is corresponding to more significant economic impact.

(AI) technology learning that automates analytical model, including predicting cause-and-effect relationship from biological data, identifying new drugs, self-driving cars and protecting against fraud etc. functions. Also (AI) learning can improve natural language processing that allows computers to continue to better analyze, understand and generate language to interface with human using the natural human language, virtual personal assistant, helps users by providing scheduling appointment, reminds organizing personal finance and finding providers of various services, machine vision allows (AI) machine man to identify object, scenes and activities in detect pedestrians and bicyclists.

We expect the economic effects of (AI) technology to include both direct GDP growth from sectors that develop or manufacture (AI) technology and indirect GDP growth through increased productivity in existing sectors that employ some from of (AI) technology. If (AI) producing sectors could grow, then it could lead to increase revenues and employment of (AI) technological professionals within these existing firms as well as the potential creation of entirely new economic activities to any countries' societies productivity improvement in existing sectors could be realized through faster and move efficient processes and decision making as well as increased (AI) technological knowledge and access to information available in societies easily.

In the future, if (AI) technology is an increasingly critical component of more products, it will become an integral part of necessary products of many people's lives. The extent of (AI)'s economy effort is also likely to vary from region to region, thought variation may be more dependent on the predominate economic activity of a region and the (AI) ability can influence economic activity, rather then the economic or developmental status of the regions. (AI) technology can move accessibility and can use source development to do international business between one country and another country.

So (AI) technology has the potential to give benefits to different income chooses and to bring significant gains to both developed and developing countries. For agricultural technology, (AI) has the potential to optimize food production around the world by analyzing agricultural regions and identifying what is necessary to improve crop yield. In total, (AI) technology gives greater economic impact to any countries agricultural regions if which implemented (AI) technology to grow crop , fruit etc. food production in the farms.

Investment in (AI) technology is such as capital investment to any countries' public or private enterprises. So, it will have large economic impact to the future . If the (AI) technology is reasonable invested to the different needs aspect by the public or private enterprises in the country. Then, it will have good economic impact to the country in the future. However, when (AI) technology is likely to affect both the productivity and employment components of economic growth in many sectors. Significant public debate has focused on projections of (AI)'s effect on the labor force. However, for instance, some researchers have argued that the rise of (AI) technology and automation will led to significant unemployment as capital is substituted for the low skillful labor. So, they point to the concern that the increasing sophistication of (AI) technology may balance skilled and semi-skilled workers and the reduce the size of the middle class. However, this is not a new argument, due to (AI) technology negatively affecting the labor force and leading to mass unemployment. Because the (AI) technology is the substitution of machinery for human labor. Although, employment in certain industries, has been reduced in the past due to technological advancement. For long term, the labor market has adapted to the introduction of new technology, giving rise to new jobs in new areas. (AI) technology may also be accomplished without a reduction to total employment in the long-term to some Asia countries, such as Hong Kong and Japan. Because Hong Kong and Japan many low skilled labor, e.g. security, cleaner who complaint that employers need them to work long time hours. (abnormal working hours) e.g. one day 12 to 15 working hour per day. Hence, if (AI) machine means invention technology success. Security or cleaning job can be worked from (AI) machine man in some hours every day in order to reduce the long time working hours cleaners or security workers, e.g. one (AI) machine man works 4 hours for cleaning or security job, one day as well as another cleaner or security labor only needs to work 8 hours one day. So total security or cleaning employers can employ 12 hours machine cleaners or security workers and human cleaners or security workers in one day. For long term benefit, Hong Kong or Japan every security or cleaning worker does not need to work 12 hours minimum working hours one day. They won't feel tried and bore and without private with whose families, so who will accept to do these cleaning or security jobs, even they can raise work efficient and performance when who feel happy and health.

So, (AI) technology of machine man invention can raise low skillful labor efficiency and it can help them to avoid abnormal working hours demand in some busy work life countries, such as Hong Kong and Japan. Before, one Japan female labor feel unhappy to work, due to who often needs to work abnormal working hours for her employer and who has less sleeping and without any private time to enjoy her life with her families every day. So this abnormal working hours factor causes her to do commit suicide behavior, then she is die unlucky. So (AI) technology of machine man invention ought avoid abnormal working hours demand for employer in any countries in the future.

The most important occurrence to any employers, some researchers had attempted to do one experiment to find that private research and development , venture capital and public research and development investment all have strong net effect or economic growth with venture capital funding further having the strongest such effect from (AI) technology. The researchers hypothesize the venture capital investment contributes to economic growth through (AI) technology innovation and by the capacity of an economy to use existing (AI) technology knowledge to increase productivity. They predict the impacts of venture capital, business-research and development and public research and development can raise multi factor productivity from (AI) technology introduction.

Can (AI) technology influence the economic development to developing countries? The developing regions of the world contain most of natural resources. If one day, (AI) technology has invent one kind of machine man which can assist any gas or oil workers to seek any new oil/gas natural resource locations easily. I believe that (AI) technology can help these natural resource exploitation countries will gain economic benefit more easily. So, (AI) driven technology can be used to change to create any new opportunities to address poor management or resources and improve human well being, such as Africa Latin America and India can use (AI) technology machine man to seek any oil/gas natural resource countries exploitation activities to attempt to gain much economic benefits.

- Can AI help offices to reduce labor number

Nowadays, increases in capital and labor are no longer driving the levels of economic growth, such as (AI) technology. The ability of increase in capital investment and in labor of traditional drivers of production, have no longer to be enjoyed in most developed economies ,e.g. developed country, US, UK . However, artificial intelligence has the potential to overcome the physical limitation of capital and labor to avoid missing out on this opportunity. So, policy makers and business leaders must prepare for and work toward a future with artificial intelligence. They must do with the idea that (AI) is another simply method to enhance productivity method . Rather they must see (AI) as the tool that can transform thinking about how growth is created.

Economists have always thought of new technologies are as driving growth their ability to enhancing. It can replace labor and capital factor of production. So, it brings this question: What is the factor of production (AI) technology characteristics. They key factor is to see (AI) technology as a capital-labor .

(AI) can replicate labor activities at much greater scale and speed, and to even perform some tasks began the capabilities of human. For example, by using virtual assistants , 1000 legal documents can be reviewed in a matter of days instead of taking three people six moths to complete. Some (AI) technology may be one kind of factor of production in the future. For another example, people will work in workplace digitalization environment. So, in the future, working environment and information management are automated. Such as Konica camera sale company will use workplace digitalization. So , (AI) technology can provide workplace digitalization in order to raise productivity efficiency. (AI) technology will be one kind of production which is replaced by workplace digitalization and it will grow any organization productivity efficiently. Then, (AI) technology will assist overall social economy growth , due to productivity is raised and products can be produced in short time to prepare to sell in consumption market. So, time will be shortened to increase GDP growth fast for the development of (AI) technology countries.

What will be the development of (AI) technology and predictions concerning the future evolution? The computers and robots will develop conscious, intelligent and minds into humans, enhancing psychological and behavioral abilities and allowing for direct communication with (AI) minds. (AI) technology will be impacted human life by (AI) technology information communicative and environmental influence. A " world brain" and " world mind", this psychological system will be enhanced and enriched the capacities of both individual and collective cognition by (AI) technology of service industries.

(AI) technology with influence these human needs of service industries changes, such as , biological science, finance, entertainment, business, biological science, transportation, communication military etc. The personal computer evolution, the internet and the world wide web which exploded on the scene, linking business, homes, schools, social organizations which were a completely unpredicted phenomenon to influence human life. Kurzweil (1999) predicts that by 2029 year, most human communication will be with machines. According to Person, by 2100 year, there will be human machine convergence.

How can (AI) technology influence environmental protection to make benefits to farming economic growth? (AI) technology can be applied to predict how to solve environmental pollution challenge to avoid to damage any crop or vegetable or rice or fruit etc. food growth. Because environmental experts can gather global environmental pollution data from an environmental database to build a perform a systematic analysis from (AI) technology. The first step is this broad analysis can include understanding, statistical and data gathering techniques to obtain the relevant data, the correlation among the variables involved, and a list of possible models. The next step is to select a set of methods and models that cover all kinds of knowledge and functionalities needed for the decision making process. Once the models are selected, they must be fully implemented by means of machine learning , data mining, statistical or numerical technique. After that, those models must be integrated to build the whole EDSS. The EDSS must be tested to check its performance, accuracy, usefulness and reliability, both from the user's and (AI) technology/ computer scientist's point of view. If these is any wrong feature in any development stage, such as model's integration, models' implementation, selection of models, database, problem analysis etc. the developers must come back in the update th required components. When the evaluation phase is all right, the EDSS is ready to be applied to the environment. The great contribution of artificial intelligence to EDSS the integration of several methods complementing the classical statistical models/simulation , statistical analysis, linear models, etc. and numerical models (control algorithms, optimization techniques etc.) . This cooperation makes the resulting systems more reliable and powerful in coping with real world environment systems. Date interpretation has been a principal area of research in (AI) technology since the very beginning. The most demanding problem in the environmental assessment context. Knowledge representation permits the definition of the different types of data that the existing methods adapt to the process. There is also a lot of work to clean, repair and transform the huge available quantities of raw data. Apart from this, the availability of meta-information or background knowledge is required to guide the process. Data mining is multi-disciplinary: It covers expert systems, data based technology, statistics, data visualization and unsupervised machine learning. These techniques operate at the level of data and background information, where numerous and often incompatible new commensurate pieces of information from disparate sources have to be brought together (K, Fedra, 1994).

So, it seems that in the future, (AI) technology with the increasing maturity in particular those related to knowledge and engineering, new dimensions can be assisted to users in environmental decision making are available. For example, many environmental systems are characterized both by incomplete models and by limited data. Hence, in the future, (AI) technology will be applied to predict climate change to reduce crop or fruit etc. food agriculture challenge by climate change bad influence.

To understand how the manufacturing business must adapt to prosper in the technology, we need to understand how (AI) technology will change us to shape our daily habits to satisfy our expectation of products to how we shop and even the immediate of the entire process. For example, taxi services are in the crosshairs as on demand transportation services like, available of the touch of a smart phone button expand. In fact, Yellow lab, US country , san Francisco city's largest taxi company is filing for bankruptcy as the industry starts to change faster than almost anyone expected. However, at this point, its more than an app that is changing, some our taxi passengers renting taxi transportation to catch consumption behavior.

(AI) technology will influence digital economy for taxi passenger's individual customer experience, offering a growing renting taxi to catch of service and feedback opportunities when any one taxi passenger who chooses to use mobile phone app online tool to prepaid to rent any taxi more easily.

Also in the long term, (AI) technology can influence vehicles drive themselves of behavior. Already, companies like Google and GM are working on projects to bring fleets of autonomous vehicles to cities at the path of a button.

Moreover, this on-demand service model is beginning to appear across a much broader range of markets. For example , Amazon company is investing in its own fleet of trucks, planes and even drone at the same time as it pushes for same-day delivery of products. As some point, vehicles will be autonomous too. So, it seems that (AI) technique will influence any transportations choose to use digital autonomous driving technology in the future . For Amazon company case, it is not stopping of logistics. It is also aiming to automatically manage the supply of consumer home products with its recently launched Amazon replenishment service, Dash. Dash is a digital service that enables that connected derive to automatically order physical products from Amazon when supplies are running low. So, it seems (AI) technology will be applied to logistic function by digital technology method introduction in the future.

Hence autonomous vehicles will optimize industry supply chains and logistics operations through increased efficiency and flexibility. In fact, fully automated and lean supply chains will keep reduce load sizes and inventory by leveraging smart distribution technologies and smaller autonomous vehicles by machine man assistance. If Amazon continues to grow market share for online sales by reducing effort required by the consumer to place an order, when also contributing the almost immediate delivery of products to the doorstep. So, it will further fuel the trend toward on-demand derive. As Amazon company fuels the on-demand economy, consumers will expect immediacy in more parts of the digital economy. On top of speed, consumers increasing expect more personalization options.

So, (AI) technology will influence digital manufacturing, such as Amazon publishing to monitor every aspect of every process in real -time and communicating to self-optimized deep learning robotics, new methods of high volume and high customization will become possible. Then, as products merge into product platforms and even services, manufacturers have the opportunity to provide components and platforms used by smaller players. So, (AI) technology will influence manufacturing industry to choose automated SMI lines, robots installed, automation engineers.

Another future (AI) technology development can be applied to space science aspect, such as Automation engineering space in manufacturing process to achieve digital manufacturing benefits to any businesses in the future. Such as reducing cost, shortening manufacturing time, raising efficiency, shortening delivery products to client individual time. How can artificial intelligence give the need and advanced fast and evaluation methods benefits for space exploration? When US NASA (space exploration organization) achieves any space exploration missions, it will answer this question:

When is it useful to have a machine use (AI) technology to achieve a decision? After all, after millions of years of space exploration and rough 10,000 years of civilization, humans are usually quite good at making decisions in complex uncertain environments. Through, Johns Hoplains University's Applied Physical Lab. Research in (AI) technology enabled systems, which has identified three general use cases for (AI) technology to explore space mission:

First, for some tasks (AI) technology is more cost effectiveness than human. Second, (AI) technology is better suited than humans at solving some, but not all problems. Third, (AI) technology allows NASA organization's space exploration mission to develop machines that ate capable of responding faster than when a human is in the decision loop (D. Scheidt, 2012, A. Castano et. al. 2008).

So, the use of (AI) technology to enable science by observing the pace of rapidly evolving phenomena was demonstrated. It is more effectively coordinating and (AI) technology utilizing to earn economic benefits to use for

space exploration mission.
However, (AI) technology also have current risk for space exploration. Today (AI) technology is immature and requires further development to reach its potential. For instance, the (AI) technology algorithms that detected the dust derive could not have identified whether the Martain weather represented a threat to the cover. Also it can not yet use instrument input to determine what, where and how to autonomously make the next space science measurement. An equally important factor limiting (AI)'s deployment is that lacks the methodology and technology to effectively test (AI) technology. So, the challenge will testing (AI) enabled system is how (AI) performance can be measured. It would be NASA organization's difficulty to find (AI) technology to develop to carry on researching any space exploration missions in the future. However, (AI) technology will be a good economic benefit choice for space exploration mission in the future.

6

Must Developed And Developing Countries Need Artificial Intelligent Development To Assist Office Tasks

Must developed and developing countries need artificial intelligent development to assist office tasks ? Ought AI is needed to prefer to develop technique to assist office staffs to reduce workload to compare other kinds of occupation environment tasks aspects ? If one developed country, e.g. US, UK , Japan , Singapore it does not continue to develop artificial intelligence, robotic, then what disadvantges or weaknesses , it will encounter to compare when it chooses to continue to develop this artificial intelligent technology in society. If one developing country, e.g. China, Korea,

Taiwan, it does not continue to develop artificial intelligence, robotic, then what disadantages or weaknesses, it will also encounter to compare when it chooses to continue to develop this artificial intelligent technology in in society. I shall explan the reasons why the results may cause to either the developed country, or the developing country as below:

- How AI help developing countries to communication and agriculture and learning and medical delivery development

Why can AI help developing countries ? Drones that pick inaccessible crops and mobile phones that give medical advice are two of the ways AI can transform life in the developing world. Artificial intelligence (AI) may improve the lives of the world's poor, the technology needed to revolutionise inefficient, ineffective food and healthcare systems in developing countries is well. For example, in low-income areas, agriculture and healthcare are two critical ecosystems that we can apply AI to immediately; this is not the far future, or even in five years.

Artificial intelligence (AI) has seeped into the daily lives of people in the developed world. From virtual assistants to recommendation engines, AI is in the news, our homes and offices. There is a lot of potential in terms of AI usage, especially in humanitarian areas. The impact could have a multiplier effect in developing countries, where resources are limited.

Emergency Response to developing countries' earthquake natural damage suddence occurrence predicting

AI and machine learning are still finding importance in emerging markets, but certain applications have emerged and are now widely used. For instance, predictive models for disaster relief enable first responders to automatically analyze large-scale behavior and movement through multiple sources of data including social media platforms, web forums, news sources, etc. Based on collected data, responders can scale reconstruction efforts and distribute supplies in a timely manner.

Why and how AI can assist farmers to predict when the earthquake occurs suddenly in order to avoid or reduce the natural damage to their agriculture productive number loss. For example, In 2015, when a major earthquake hit Nepal, more than 8 million people were affected. During the aftermath, drones were used to map and assess the destruction and speed up the rescue mission. The town of Sankhu, situated about 20 kilometers northeast of Kathmandu, was among the highly affected locations. In May 2018, my company Fusemachines and GeoSpatial Systems partnered with Sankhu's city officials to use drones and artificial intelligence in an effort to automatically estimate the reconstruction need. After processing data accumulated from a drone-powered aerial mapping of the region, the team fed this data to advanced machine learning algorithms. Combining drone imagery, digital mapping and machine learning, the team configured region modeling and infrastructure development with higher accuracy. Another organization known as One Concern, a California-based startup, has created a predictive AI program called Seismic Concern to accurately predict seism and is also working on solutions for wildfires, floods and hurricanes.

Smart AI Agriculture

Another application of AI in developing countries is smart agriculture. Farmers monitor crops more effectively and make better predictions on planting, weeding and harvesting using AI tools. It can also be used to analyze one plant at a time and add pesticides only to infected plants and trees instead of spraying pesticides across large swaths of crops. One California-based tech company is an example of this use of AI. So, the developing countries farmers in rural parts of India are also using AI to increase yields through better access to information about the farming season than they would normally have. Technology-enabled process automation offers the agribusiness industry the chance for remarkable growth -- not only in developed countries but around the world. There's a unique opportunity to increase yields, cut down labor costs and improve people's health.

Medicine Delivery to developing countries' patients urgent need

Companies are also leveraging AI to improve access to health care in some of the most remote areas of the world. In Rwanda, for example, Zipline is using drones to deliver medical supplies and blood to hospitals and clinics that are difficult to access by car. This has dramatically impacted people living in remote parts of the country because they are able to get medical help when needed. The drone system in Rwanda has also helped reduce waste of blood by 95%, as noted by Zipline. One Concern has created an AI program called Seismic Concern that accurately predicts seismic events and is also working on solutions for floods, wildfires and hurricanes. The medical field may actually benefit the most from emerging technologies in developing countries.

Assistance to reduce teaching work workload or psychological pressure to teachers in developing countries' schools

Another vital area benefiting from innovative technologies like AI is education. Advanced technologies can enhance how we learn, teach and perform tasks. In most developing countries, schools lack experienced teachers and resources to enhance students' knowledge. As a result, many students still have to walk long distances to get to the nearest school, which has created education gaps, especially in rural areas. AI tools such as personalized learning assistants can simplify learning by making tutoring services and learning materials accessible to all students, wherever they are. Machines can be automated to help students learn basic concepts without a tutor, which companies like Carnegie Learning are working on. This would allow students to learn at any time from anywhere. With AI, education is made easy and accessible to more people.

The initial usage of AI in developing countries has been at a micro level -- solving small, specific problems in a defined industry. As machine learning advances and there is a higher utilization of AI, we will see more complex issues being targeted and resolved. When duly adopted, AI can positively impact future developing countries people everyday lives not just in disaster intervention, education, health care and agriculture but can also help in mitigating poverty, malnutrition and pollution. Especially, in developing nations, to leverage AI's true potential and create a snowball effect. Startups are defining a holistic and humanitarian approach to building more sophisticated, AI-ready societies. Stakeholders in the AI landscape should understand the strengths and nuances of the developing world as well as the limitations of AI and create localized solutions and applications.

Why does smart phone help developing countries communication ?
Internet Seen as Positive Influence on Education but Negative on Morality in Emerging and Developing Nations. Internet access differs substantially across the 32 emerging and developing countries polled, with the lowest rates of internet use in South Asian and sub-Saharan African nations. Within countries, computer owners, young people, the well-educated, the wealthy and those with English language ability are much more likely to access the internet than their counterparts. To access the internet, people increasingly use smartphones rather than more cumbersome fixed landline connections and computers. Around the world, both smartphones and basic-feature phones alike are used for sending messages and taking pictures.
In fact, many developing countries young people, students are popular to use smart phones for internet usage aim, instead of communication. Moreover, many developing countries working people are also popular to use smart phones for any working usage in their working time , even non working time any time. So, smart phones (AI) phones will be important communication or leisure tools to developing countries people in the future. Unless, it is one day, scientists can develop another new communication tool to replace smart phones. So, artificial intelligence will be important to influence developing countries people , how to improve or bring positive learning attitudes to students in their daily learnnng lifes. as well as how to raise developing countries people, how to raise working people efficiency or improve performace in their daily working lifes. So, AI may bring positive learning or working attitudes to developing countries working people and

students both.

The Positive Impact of Mass Media in Developing Countries

Radio, newspapers, television, Internet, social media, etc., all of these are forms of mass media. Each of these outlets has the capability of bringing information to thousands of people with one device. While in some communities it is easy to take advantage of these communication outlets such as television and Internet access, not everyone has access to such outlets. Radio is one of the most common forms of mass media in developing countries because it's affordable and uses less electricity than many other forms of mass media, but only approximately 75 percent of people in developing countries have access to a radio, and roughly 77 percent of people in rural areas have access to electricity. For developing countries that have implemented forms of mass media in their communities, there have been numerous positive outcomes are influenced to impact developing countries mass media by artificial intelligence as below:

When AI is participated to developing countries mass media, it can influence any radio, television audiences raise more attention to each other through social media platforms such as Facebook and Twitter and create, organize and initiate street protests and campaigns. Furthermore, having access to social media in developing countries, people are able to connect to those that they usually wouldn't have the chance to talk to. Moreover, AI Provides educational opportunities- In many countries, the division between local and national languages as well as issues of literacy can make communication difficult. With the use of mass media, a bridge can be built between these two gaps. In India, there is a radio station that provides information in local languages and respects local culture and traditions. One of the main ways is to create public awareness of what is going on with businesses and government officials. The media plays an important role in giving people the opportunity to act against injustice, oppression and misdeeds that they otherwise wouldn't know about. Information on available healthcare, a mass radio broadcast was sent out encouraging parents to seek treatment at local healthcare facilities for their sick children. With this mass outreach on healthcare, the encouragement of people to take their children to healthcare facilities saved thousands of lives. This easy way of encouraging others and bringing awareness about certain diseases was made possible through a simple radio broadcast. Finally, when AI is particiapted to media, it may bring many social issues to life that otherwise would remain unknown to many people. In developing countries and communities like Burkina Faso, when the radio broadcast was released about malaria, diarrhea and pneumonia, people were educated and moved to action and knew to take their children to healthcare facilities for preventative care. As it is seen, having access to different media outlets is vital for those in developing countries. Here are three ways that those in developing countries can implement mass media to help their people and communities.

When AI is participated to any internet radio or internet newspaper mass online listening or reading channel. It can provide online radios or newspapers in public places- By providing online radios and newspapers in public areas it gives community members to access news, information and emergency warnings. Even though radios can be on the cheaper side, there are still many people that can't afford to have a radio in their home. By providing one in a local place, not only would it better educate the community members but also it will bring the community together. So, it can make media outlets a two-way platform- Creating a two-way platform between the community and those who are behind the radio stations, newspapers or broadcasts makes the community feel involved and that their voices are being heard. An organization called Soul City in sub-Saharan Africa is showing how well two-way platforms work by engaging their listeners and having them contribute thoughts and ideas about complex issues. Because developing countries radio listening audiences or newspaper readers are popular to accept computer online radio listening channel or online newspaper reading channel to replace traditional paper newspapers or radio machines. So, AI may raise their listening news or reading news leisure feeling from online mass media channel in the future.

- Why do developed countries need to develop AI

Artificial intelligence, or AI, is driving massive shifts across the globe, and every day more questions arise. What impact will AI have on the workforce and how can we prepare for it? How can we encourage economy-boosting and job-creating technologies? How can we ensure that AI will be implemented ethically and with minimal bias? How will society benefit? For developed country, such as US example. None of the US, Israel and Russia have a formal national AI policy yet. Private sector companies such as Google, Amazon and Apple and the US department of defence are driving the bulk of AI investment in the United States. Though Israel does not have a specific policy, it is keenly focused on AI and has seen the number of AI start-ups triple

since 2014.
Developed country may learn whether what weakness it is lacking when it does not continue to develop AI from one another developed country. Which countries are approaching AI most effectively, and to what degree is there opportunity for greater international collaboration? It may be too early to tell; however, when analyzing the best practices of existing national AI policies, there is much that can be learned. These are the specific areas to consider. When one developed country continue to develop or research AI, it may bring these benefits as below:
On gathering Data aspect, from self-driving vehicles to smart cities, data is the driver behind AI. Innovation in the United States is limited without a national strategy that answers questions about protocol and ownership. France and Denmark, on the other hand, are opening government data. France is hosting troves of centrally collected public and private data that it plans to make available as part of its strategy. Conversely, by taking a restrictive position on issues of data collection (as indicated by the implementation of General Data Protection Regulation), the EU is putting manufacturers and software designers at a disadvantage while balancing the demand for privacy. On raising technologica talent aspect, the demand for AI talent far outweighs the available supply. As a result, almost every nation's strategy addresses talent development. Canada's AI strategy is distinct in that it primarily focuses on research and talent strategy. The country boasts AI degree programmes and is building a $127 million research facility in Toronto. Companies like Facebook and my own company, Uptake, are investing in Canada to access this talent pool. On AI legal technological innovation aspect, a whole host of legal questions swirl around AI. The country

is developing a bill for AI liability that will be ready in March 2019. The government hopes the legal framework will attract investors by providing a simple, comprehensive guideline to enable the broad use of AI systems. So, when the developed country applied AI technology to assist any lawyers to work, then AI can help them to reduce the workload to draft any legal documents more easier. So, any developed countries lawyers' draft legal documents time must reduce if the developed countries lawyers accept to apply AI to assist their legal works. One of the great promises of AI is its potential for improving quality of life. But without the right planning and oversight, we risk exacerbating problems of inequality or marginalizing groups of people. As an example, India's AI strategy is focused on leveraging the technology not only for economic growth, but also for social inclusion.

AI may bring what benefits to developed countries

From SIRI to self-driving cars, artificial intelligence (AI) is progressing rapidly. While science fiction often portrays AI as robots with human-like characteristics, AI can encompass anything from Google's search algorithms to IBM's Watson to autonomous weapons. Artificial intelligence today is properly known as narrow AI (or weak AI), in that it is designed to perform a narrow task (e.g. only facial recognition or only internet searches or only driving a car). However, the long-term goal of many researchers is to create general AI (AGI or strong AI). While narrow AI may outperform humans at whatever its specific task is, like playing chess or solving equations, AGI would outperform humans at nearly every cognitive task.

Why research AI safety? Would AI bring war when AI is continued to develop by developed countries? In the near term, the goal of keeping AI's impact on society beneficial motivates research in many areas, from economics and law to technical topics such as verification, validity, security and control. Whereas it may be little more than a minor nuisance if your laptop crashes or gets hacked, it becomes all the more important that an AI system does what you want it to do if it controls your car, your airplane, your pacemaker, your automated trading system or your power grid. Another short-term challenge is preventing a devastating arms race in lethal autonomous weapons.

In the long term, an important question is what will happen if the quest for strong AI succeeds and an AI system becomes better than humans at all cognitive tasks. As pointed out by I.J. Good in 1965, designing smarter AI systems is itself a cognitive task. Such a system could potentially undergo recursive self-improvement, triggering an intelligence explosion leaving human intellect far behind. By inventing revolutionary new technologies, such a superintelligence might help us eradicate war, disease, and poverty, and so the creation of strong AI might be the biggest event in human history. Some experts have expressed concern, though, that it might also be the last, unless we learn to align the goals of the AI with ours before it becomes superintelligent.

There are some who question whether strong AI will ever be achieved, and others who insist that the creation of superintelligent AI is guaranteed to be beneficial. At FLI we recognize both of these possibilities, but also recognize the potential for an artificial intelligence system to intentionally or unintentionally cause great harm. We believe research today will help us better prepare for and prevent such potentially negative consequences in the future, thus enjoying the benefits of AI while avoiding pitfalls.

How can AI be dangerous when developed countries continue to develop AI to become weapon to replace soldiers?

Most researchers agree that a superintelligent AI is unlikely to exhibit human emotions like love or hate, and that there is no reason to expect AI to become intentionally benevolent or malevolent. Instead, when considering how AI might become a risk, experts think two scenarios most likely:

The AI is programmed to do something devastating: Autonomous weapons are artificial intelligence systems that are programmed to kill. In the hands of the wrong person, these weapons could easily cause mass casualties. Moreover, an AI arms race could inadvertently lead to an AI war that also results in mass casualties. To avoid being thwarted by the enemy, these weapons would be designed to be extremely difficult to simply "turn off," so humans could plausibly lose control of such a situation. This risk is one that's present even with narrow AI, but grows as levels of AI intelligence and autonomy increase.

The AI is programmed to do something beneficial, but it develops a destructive method for achieving its goal: This can happen whenever we fail to fully align the AI's goals with ours, which is strikingly difficult. If you ask an obedient intelligent car to take you to the airport as fast as possible, it might get you there chased by helicopters and covered in vomit, doing not what you wanted but literally what you asked for. If a superintelligent system is tasked with a ambitious geoengineering project, it might wreak havoc with our ecosystem as a side effect, and view human attempts to stop it as a threat to be met. So, a super-intelligent AI will be extremely good at accomplishing its goals, and if those goals aren't aligned with ours, we have a problem. You're probably not an evil ant-hater who steps on ants out of malice, but if you're in charge of a hydroelectric green energy project and there's an anthill in the region to be flooded, too bad for the ants. A key goal of AI safety research is to never place humanity in the position of those ants.

Why the recent interest in AI safety ?

Stephen Hawking, Elon Musk, Steve Wozniak, Bill Gates, and many other big names in science and technology have recently expressed concern in the media and via open letters about the risks posed by AI, joined by many leading AI researchers. The idea that the quest for strong AI would ultimately succeed was long thought of as science fiction, centuries or more away. However, thanks to recent breakthroughs, many AI milestones, which experts viewed as decades away merely five years ago, have now been reached, making many experts take seriously the possibility of superintelligence in our lifetime. While some experts still guess that human-level AI is centuries away, most AI researches at the 2015 Puerto Rico Conference guessed that it would happen before 2060. Since it may take decades to complete the required safety research, it is prudent to start it now.

Because AI has the potential to become more intelligent than any human, we have no surprise way of predicting how it will behave. We can't use past technological developments as much of a basis because we've never created anything that has the ability to, wittingly or unwittingly, outsmart us. The best example of what we could face may be our own evolution. People now control the planet, not because we're the strongest, fastest or biggest, but because we're the smartest. If we're no longer the smartest, are we assured to remain in control?

A captivating conversation is taking place about the future of artificial intelligence and what it will/should mean for humanity. There are fascinating controversies where the world's leading experts disagree, such as: AI's future impact on the job market; if/when human-level AI will be developed; whether this will lead to an intelligence explosion; and whether this is something we should welcome or fear. But there are also many examples of of boring pseudo-controversies caused by people misunderstanding and talking past each other. When one developed country continue to develop AI, can itself country's all factories workers will lose jobs, due to AI can replace them to do simple works in factories, or any public transport drivers, e.g. bus drivers, ferry , tram, train drivers, they will lose jobs, when AI (non manual driving drivers) can replace all public transport drivers. So, some occupations will lose if developed countries continue to develop or research AI to replace human to do some simple jobs, such as some cooking jobs can be done by AI. So, it is possible that future cookers won't be needed, because AI cooking skills may be better than them to cook any good taste chinese or western food in restaurants. If you drive down the road, you have a subjective experience of colors, sounds, etc. But does a self-driving car have a subjective experience? Does it feel like anything at all to be a self-driving car? Although this mystery of consciousness is interesting in its own right, it's irrelevant to AI risk. If you get struck by a driverless car, it makes no difference to you whether it subjectively feels conscious. In the same way, what will affect us humans is what superintelligent AI does, not how it subjectively feels.

In fact, AI may be make any brokers jobs in financial market. the main concern of the beneficial-AI movement isn't with robots but with intelligence itself: specifically, intelligence whose goals are misaligned with ours. To cause us trouble, such misaligned superhuman intelligence needs no robotic body, merely an internet connection – this may enable outsmarting financial markets, out-inventing human researchers, out-manipulating human leaders, and developing weapons we cannot even understand. Even if building robots were physically impossible, a super-intelligent and super-wealthy AI could easily pay or manipulate many humans to unwittingly do its bidding. So, future brokers will be replaced by AI, when AI can be made to own financial brokers' analytical mind to make more accurate whether the share price will rise up or fall down to compare human financial brokers' analytical mind. The robot misconception is related to the myth that machines can't control humans. Intelligence enables control: humans control tigers not because we are stronger, but because we are smarter. This means that if we cede our position as smartest on our planet, it's possible that we might also cede control.

Not wasting time on the above-mentioned misconceptions lets us focus on true and interesting controversies where even the experts disagree. What sort of future do you want? Should we develop lethal autonomous weapons? What would you like to happen with job automation? What career advice would you give today's kids? Do you prefer new jobs replacing the old ones, or a jobless society where everyone enjoys a life of leisure and machine-produced wealth? Further down the road, would you like us to create superintelligent life and spread it through our cosmos? Will we control intelligent machines or will they control us? Will intelligent machines replace us, coexist with us, or merge with us? What will it mean to be human in the age of artificial intelligence?

Why do developed countries people need AI ?

Why do we assume that AI will require more and more physical space and more power when human intelligence continuously manages to miniaturize and reduce power consumption of its devices. How low the power needs and how small will the machines be by the time quantum computing becomes reality? Why do we assume that AI will exist as independent machines? If so, and the AI is able to improve its Intelligence by reprogramming itself, will machines driven by slower processors feel threatened, not by mere stupid humans, but by machines with faster processors? What would drive machines to reproduce themselves when there is no biological incentive, pressure or need to do so?

Who says superior AI will need or want to have a physical existence when an immaterial AI could evolve and preserve itself better from external dangers. What will happen if AI developed by competing ideologies, liberalism vs communism, reach maturity at the same time, will they fight for hegemony by trying to destroy each other physically and/or virtually. If AI is programmed to believe in God, and competing AI emerges programmed by muslims, christians or jews, how are the different AI's going to make sense of the different religious beliefs, are we going to have AI religious wars? What if the "powers that be" greatest fear is the emergence of a super AI that police's and rationalizes the distribution of wealth and food. A friendly super AI that is programmed to help humanity by, enforcing the declaration of Human Rights (the US is the only industrialized country that to this day has not signed this declaration) ending corruption and racism and protecting the environment.Most benefits of civilization stem from intelligence, so how can we enhance these benefits with artificial intelligence without being replaced on the job market and perhaps altogether?

Key to the process of machine learning are neural networks. These are brain-inspired networks of interconnected layers of algorithms, called neurons, that feed data into each other, and which can be trained to carry out specific tasks by modifying the importance attributed to input data as it passes between the layers. During training of these neural networks, the weights attached to different inputs will continue to be varied until the output from the neural network is very close to what is desired, at which point the network will have 'learned' how to carry out a particular task. A subset of machine learning is deep learning, where neural networks are expanded into sprawling networks with a huge number of layers that are trained using massive amounts of data. It is these deep neural networks that have fuelled the current leap forward in the ability of computers to carry out task like speech recognition and computer vision.

In conclusion, when developed countries continue to develop AI, it may bring positive advantages to bring raising productivies, or efficiencies, but it may also raise unemployment ratio to any low skill or low knowledge jobs in ther societies. However, human future society will need to change to be better to raise our living standard. But AI is one kind the best choice tool to achieve this aim in our future, so I agree developed countries continue to develop or research AI to be the super -human machine.

Printed by Libri Plureos GmbH in Hamburg, Germany